I0001008

Orson Pratt

New and Easy Method of Solution of the Cubic and Biquadratic

Equations

Orson Pratt

New and Easy Method of Solution of the Cubic and Biquadratic Equations

ISBN/EAN: 9783337389734

Printed in Europe, USA, Canada, Australia, Japan

Cover: Foto ©berggeist007 / pixelio.de

More available books at **www.hansebooks.com**

NEW AND EASY METHOD

OF SOLUTION OF THE

CUBIC AND BIQUADRATIC EQUATIONS,

EMBRACING SEVERAL NEW FORMULAS,

GREATLY SIMPLIFYING THIS DEPARTMENT OF MATHEMATICAL SCIENCE.

DESIGNED AS

A SEQUEL TO THE ELEMENTS OF ALGEBRA,

AND FOR

THE USE OF SCHOOLS AND ACADEMIES.

By ORSON PRATT, Sen.

LONDON:
LONGMANS, GREEN, READER, AND DYER.
LIVERPOOL:
B. YOUNG Jun., 42, ISLINGTON.

Right of Translation reserved.
MAY, 1866.

ENTERED AT STATIONERS' HALL BY THE AUTHOR.

D. MARPLES, PRINTER, LIVERPOOL.

PREFACE.

THE beauty of any scientific theory is simplicity : this has been the aim of the Author in the composition of the following pages. He has sought to render the Solution of Equations of the Third and Fourth Degrees in a more simple form, adapted to an elementary course of instruction in Algebra, to be used in schools and academies.

In most treatises on common Algebra the solution and theory of Quadratic equations, or equations of the second degree, are clearly developed in formula simple and easy to be comprehended, and generally accompanied with numerous examples, calculated to interest and encourage the young student; but Equations of the Third and Fourth Degrees, though often introduced by algebraical writers, have been presented in such an unfavourable aspect, and encumbered with so many complex rules, requiring such a vast amount of labour, that the pupil, instead of being interested, becomes weary, and often disgusted, with the obscurity of this department of his subject.

Other writers, celebrated in the annals of mathematical science, perceiving the great disadvantages resulting from the incorporation of these formidable obstacles among the rudiments of Algebra, have excluded them from these elements, and formed separate treatises, having for their object the analyzation and solution of these two orders of equations: but, instead of discovering some simple method, by which the roots can be at once numerically obtained, numerous pages, in these volumes, are appropriated in discussing the limits of the roots, and in a preparatory analysis by the theorem of STURM, or some other laborious theorem, which is considered by them as essential, before the solution by the tedious method of HORNER, or by some other similar method, can commence.

These cumbersome theories, in certain cases so unsatisfactory in their results, induced the Author to carefully investigate the nature and properties of the Cubic and Biquadratic Equations, to discover, if possible, some simpler method of solution, which should supersede the complicated processes adverted to above, and bring these two equations in closer alliance to the Quadratic, rendering them suitable to be incorporated in all elementary treatises of Algebra. How far he has succeeded must be left for the following pages and the judgment of mathematicians to determine. It is, however, but justice to the public that a brief statement of the superiority of this new method should be set forth in these prefatory remarks.

1. The Author's discovery of the *Equation of Differences*, together with several other kindred discoveries, resulting from the properties of this equation, has enabled him to entirely dispense with every process for finding the limits of the roots; to dispense with the theorem of STURM, and all similar theorems, having for their object the determination of the number of real roots and their situation in the arithmetical scale; to dispense with all processes for finding the first figure of a root by trial or successive substitution; and to dispense with the successive trial divisors used by HORNER.

2. By this new method the first figure of a root is found in the same manner that the first figure of any quotient is

obtained; and each new divisor is derived, by a simple formula, from the figures of the root already developed: and instead of each divisor being disjointed from the dividend, and placed in a column far distant, as is the case in HORNER's method, it is made to occupy its usual place on the left of the dividend, as in common division; thus reducing the whole process into a more compact and simple form, more in accordance with the usual arithmetical form of extracting the square root, which it, in some respects, resembles.

3. All those cases in which the roots approximate each other in value, hitherto considered so difficult of solution, become, by this method, exceedingly simple; indeed, the nearer two roots approach equality the less is the labour in the operation of development.

4. A new process, simple and expeditious, has been devised for obtaining the remaining roots of a Cubic or Biquadratic equation, after one root has been found, with-out resorting to the common, or more tedious, method of depressing the equation.

5. A *new general formula* has been discovered, by which the three roots of a Cubic equation, when they are all *real*, can be obtained in terms of the co-efficients, without re-sorting to the process of development figure by figure.

6. A new and simple method of extracting the cube root is given, by which the labour becomes several times less than by the usual methods. This very expeditious process requires only about the same number of figures as extracting the square root, and constantly maintains the divisors in the same horizontal lines with their respective dividends.

7. General Cubic and Biquadratic Equations which have, in all cases, two equal roots, are given, and considered by the Author of considerable importance in their relative bearings upon other equations.

8. A *"General Solution"* of the Biquadratic Equation is given, resembling in some respects DESCARTES' Solution, but differing in other respects from all solutions with which the Author is acquainted, by obtaining a resulting auxiliary Cubic Equation whose second term is absent.

These are some of the peculiarities in this little treatise ; but the reader is referred to the propositions in the body of the work for further information.

In the meantime, the Author begs the indulgence of the public for obtruding upon them new discoveries, new theorems, and new formulas, calculated to weaken the old methods of instruction which, through age, are so highly venerated among the learned institutions of civilised nations. The Author makes no pretensions to literary merit, being *" self taught,"* and has composed the following propositions under very unfavourable circumstances, in the midst of the bustling and exciting sceneries of a continental tour in Europe, without access to books and libraries to which he could refer on many points of importance ; his style and arrangement will, therefore, undoubtedly appear very imperfect, and open to severe criticism.

But should the Author, in his humble capacity, succeed in contributing even *one new truth* to the enlargement of the sphere of mathematical knowledge, or be instrumental in simplifying any department of this useful science, so as to render it easier and more accessible to the general student, he will have attained the desirable object he had in anticipation.

ORSON PRATT, Sen.

VIENNA, AUSTRIA, *August*, 1865.

REMARKS.

Since his return from the Continent, the Author has introduced several improvements and simplifications in the numerical department of this treatise, not specified in the foregoing preface; among which may be mentioned the application of his new theory to the extraction of the Biquadratic Root, by which a great amount of labour connected with the old methods is avoided, and the development rendered in an abbreviated style, extremely simple and expeditious. That the work might be more fully adapted to the wants of every class of algebraical students, either with or without teachers, the Author has introduced a great variety of examples, with numerical operations, indicating the most convenient form of arrangement to be pursued. With these few remarks, he submits, with feelings of pleasure, mingled with diffidence, this little volume, as a humble contribution to the great treasury of mathematical science.

O. PRATT, Sen.

42, Islington, Liverpool, *May* 26, 1866.

ERRATA.

PAGE 6, nine lines from top, *for* $\left(x - a_3 \right.$ *read* $\left(x - a_3\right)$.

— 29, four lines from top, *for* A_0^3 *read* A_0^2.

— 35, seven lines from bottom, *for* Y *read* $Y_,$.

— 35, last line, *for* $\sqrt{} \pm$ *read* $\pm \sqrt{}$.

— 38, two lines from bottom, *for* $- 3\,A + A_2^2$ *read* $- 3\,A + A_2^2$.

— 76, three lines from top, *for* $- a$ *read* a.

— 103, example 7, middle column, the group of figures, below the seventh row, is displaced two figures to the right.

— 104, ten lines from top, *for* $- y''$ *read* $- y'''$.

— 110, Example 11, third column,

in row 2, *for* 60 *read* 96;

„   5, *for* 12 *read* 13;

„   6, *for* 71 *read* 78;

„   12, *for* ·01 *read* ·08;

„   17, *for* ·03 *read* ·00.

CONTENTS.

INTRODUCTION.

CHAPTER I.

CONSTITUTION OF CUBIC AND BIQUADRATIC EQUATIONS.

CHAPTER II.

TRANSFORMATION OF CUBIC AND BIQUADRATIC EQUATIONS.

CHAPTER III.

REAL AND IMAGINARY ROOTS.

CHAPTER IV.

EQUAL ROOTS.

CHAPTER V.

CUBE ROOTS.

CHAPTER VII.

BIQUADRATIC EQUATIONS.

CHAPTER IX.

NUMERICAL SOLUTION OF BIQUADRATIC EQUATIONS.

SOLUTION OF EQUATIONS

OF THE

THIRD AND FOURTH DEGREES.

INTRODUCTION.

(ART. 1.) All those preliminary investigations, so essential to the general theory of equations of a higher order than that of the fourth degree, are, if unnecessary to the solution of the cubic and biquadratic equations, carefully excluded; so as not to encumber the simplicity of our method with details or theorems more properly belonging to a higher department. The theory of quadratic equations, or equations of the second degree, is likewise excluded; as the student is supposed to have made himself familiar with this class of equations, so amply elucidated in the elements of common Algebra.

(2.) The definition of a cubic equation is an equation in which the highest power of the unknown quantity is a cube; and in which the inferior powers of the unknown quantity are neither fractional nor negative, but integral; and whose coefficients are known quantities, either real or imaginary, fractional or integral; and the sum of whose terms, when removed to one side of the sign of equality, is equal to nothing: thus

$$A_3 x^3 + A_2 x^2 + A x + A_0 = 0$$

is a *complete* cubic equation: if one or more of the last three terms are absent, the equation is called *incomplete*. The coefficients A_3, A_2, A, A_0 may be either positive or negative: the

1

sign $+$, used in this general equation, is merely to connect the terms together, but indicates nothing further. A biquadratic equation is one in which the highest power of the unknown quantity is four ; but, in other respects, the definitions regarding the inferior powers of the unknown quantity, and the coefficients are the same as in the cubic : thus :

$$A_4\,x^4 + A_3\,x^3 + A_2\,x^2 + Ax + A_0 = 0$$

is a complete biquadratic equation.

(3.) When these equations involve but one unknown quantity, they are called *determinate* equations ; but when two or more unknown quantities are involved, they are called *indeterminate* equations. Certain groups of equations, involving more than one unknown quantity, may be reduced to as many determinate equations as there are unknown quantities, providing that there are as many distinct and independent equations as there are unknown quantities; otherwise they are indeterminate. It is sometimes the case, when one of the powers of the unknown quantity is negative or fractional, that the equation can be reduced, by certain algebraical transformations, to an integral form ; for example, the equation $ax^3 + bx^2 + cx + dx^{-1} + e = 0$ may be transformed into $ax^4 + bx^3 + cx^2 + ex + d = 0$; also the equation $ax^2 + b\,x + cx^{\frac{1}{2}} + d = 0$ may be reduced, by transposing the third term and squaring both sides of the equation, to a rational integral equation of the fourth degree. Rational integral determinate equations of the third and fourth degrees, are the only ones which will be considered in the following pages.

(4.) The general equation can be more simplified by reducing the coefficient of the highest power of the unknown quantity to unity : this can be effected by dividing each term by that coefficient, which of course will not alter the value of the equation : thus

$$x^4 + \frac{A_3}{A_4}\,x^3 + \frac{A_2}{A_4}\,x^2 + \frac{A}{A_4}\,x + \frac{A_0}{A_4} = 0$$

that is $x^4 + A'_3\,x^3 + A'_2\,x^2 + A'\,x + A'_0 = 0$

This form will be adopted because of its greater simplicity. The accents may also be dispensed with. For the convenience of reference, the left hand member of the general equation will be represented by the capital letter X. The two equations in this modified simple form will stand as follows:

$$X = x^3 + A_2 x^2 + Ax + A_0 = 0.$$

$$X = x^4 + A_3 x^3 + A_2 x^2 + Ax + A_0 = 0.$$

(5.) If all the terms of an equation are arranged on one side of the sign of equality, and of course equal to nothing, then any quantity which, when substituted for x, does not alter the value of the equation, but maintains it in its identity to zero, is a *root* of the equation: this can be more simply explained by a few examples. If $x - a = 0$ and a is substituted for x, the identity of the first member of the equation to zero is not altered; therefore a is a root of this simple equation; and it is very easy to perceive that the substitution of any other value for x but that of a will destroy the equality of the first member to zero: and therefore, the equation can have only one root. If $x^2 - a^2 = 0$, and either $+a$ or $-a$ be substituted for x, the identity of the first member to zero is still preserved; and therefore, both $+a$ and $-a$ are roots of this equation. And it is also proved in common Algebra that the equation $ax^2 + bx + c = 0$ has two roots, namely, $\dfrac{-b + \sqrt{b^2 - 4ac}}{2a}$ and $\dfrac{-b - \sqrt{b^2 - 4ac}}{2a}$, either of which when substituted for the unknown quantity in the equation preserves its identity to zero. In the subsequent pages of this treatise it will be proved that the cubic equation has *three roots*, and that the biquadratic equation has *four roots*, either of which, when substituted for the unknown quantity, will not disturb the conditions of the equation in respect to its zero value.

(6.) The *solution* of an equation is to determine the value of all its roots. There are two kinds of solution; one is to find the value of the roots in terms of the coefficients, when represented by symbols, which is called an *algebraical* solution: the other is to determine the numerical value of the roots, or their value in figures; this is called a *numerical* solution.

Algebraical solutions may be divided into two classes.

1. When the symbols, representing the coefficients, are obtained in a form that is susceptible of being reduced to numerical calculation, so that the values of the roots can be numerically obtained.

2. When the solution is expressed in known symbols, obtained in a form which is irreducible by any known process.

It will be seen hereafter, that what is commonly called the general solution of the cubic and biquadratic equations, is merely a transformation of the unknown quantities into known symbols, though expressed, in case of real roots, in forms that are unknown and irreducible; the latter is, therefore, erroneously called a *solution*: it would have been much more consistent to have called it a *transformation*. Further remarks will be made upon this subject when we come to treat upon the general solution of the cubic equation by CARDAN. In the meantime, we shall continue the use of the term "general solution," according to the established practice among mathematicians.

(7.) We might here introduce those general solutions, and thus demonstrate what was adverted to in article (5), namely, that the cubic equation has three roots, and the biquadratic equation four roots: but to do this would interrupt that simplicity of arrangement which is so desirable to be preserved. This property of those equations will, therefore, be assumed in some of the propositions which follow, which will merely have a tendency to render the demonstrations dependent upon this property hypothetical, until the student shall arrive at the general solutions referred to.

CHAPTER I.

PROPOSITION I.

(8.) If a_1, a_2, a_3, a_4 are roots of the biquadratic equation,
$$x^4 + A_3 x^3 + A_2 x^2 + Ax + A_0 = 0,$$
then will
$$x^4 + A_3 x^3 + A_2 x^2 + Ax + A_0 = (x - a_1)(x - a_2)(x - a_3)(x - a_4) = 0.$$

Demonstration.—As a_1, a_2, a_3, a_4 are roots of the equation, x must have these four values, and consequently be equal to them; hence each of the factors, $(x - a_1)$, $(x - a_2)$, &c., must be equal to nothing; and therefore, their product must be equal to the first member of the equation. The same proposition is also true for the cubic equation.

COROLLARY.

(9.) If $X = 0$ is divided by any one of the factors in the second member of the equation, the quotient will be equal to the product of the remaining three factors, which is also equal to nothing. And if $X = 0$ is divided by the product of any two of the factors, the quotient will be equal to the product of the remaining two factors, which is likewise equal to nothing. And if $X = 0$ is divided by the product of any three of the factors, the quotient will be the remaining factor, which is also equal to nothing. And in all these divisions the remainder will be zero. The common rules of algebraical division demonstrate this corollary.

(10.) It will be perceived by this corollary that when one root of $X = 0$ is known, as, for example, $x = a_1$, that by transposing a_1 to the same side with x, and changing the sign, we obtain a factor

$x - a_1$, which will divide the polynomial $X = 0$, and thus depress the equation to one degree lower than the original equation.

PROPOSITION II.

(11.) If $X = 0$ is any equation of the third or fourth degree, the number of roots in the equation cannot exceed the highest power of the unknown quantity.

Demonstration.—It has been proved in proposition I., that X can be exhibited in the form of factors, each equal to zero ; thus

$$X = (x - a_1)\,(x - a_2)\,(x - a_3)(x - a_4)$$

If X is depressed to an equation of the first degree by a successive division of these factors, it is evident that it will not admit of any further depression by any new factor of the form of $x - a_5$, without destroying its equality to zero ; therefore, the number of roots in the equation $X = 0$ cannot exceed the highest power of the unknown quantity.

PROPOSITION III.

(12.) If $X = 0$ is any equation of the third or fourth degree, and is divisible by any factor $x - a$, then will a be a root of the equation.

Demonstration.—Let $X = (x - a_1)\,(x - a_2)\,(x - a_3)\,(x - a_4)$; a_1, a_2, a_3, a_4 being roots of X. It is proved in proposition II., that the number of factors in X of the form of $x - a$ can never exceed the highest power of the unknown quantity ; consequently, if $x - a$ divides X, it must be equal to one of the factors in the right hand member of the equation ; and therefore a must be equal to one of the roots a_1, a_2, a_3, a_4.

PROPOSITION IV.

(13.) If any rational integral polynomial

$$X = x^4 + A_3\,x^3 + A_2\,x^2 + A\,x + A_0$$

is divided by any binomial of the form of $x - a$, the resulting remainder will be

$$a^4 + A_3\, a^3 + A_2\, a^2 + A\, a + A_0.$$

Demonstration.—Let the remainder be represented by R, and the quotient by Q ; then we shall have

$$x^4 + A_3\, x^3 + A_2\, x^2 + A\, x + A_0 = (x - a)\, Q + R;$$

now, if a is substituted for x, the first term of the second member will vanish ; and, therefore,

$$a^4 + A_3\, a^3 + A_2\, a^2 + A\, a + A_0 = R.$$

This proposition, like those which have preceded, might, with a very little alteration, be rendered general, so as to be applicable to polynomials of any degree : but, for the sake of beginners, equations and polynomials of the nth degree have been excluded, and the propositions have been confined to those of the lower orders. The young student will, however, at once perceive, from the general nature of the demonstrations, that a similar process is also applicable to polynomials and equations of any degree.

PROPOSITION V.

(14.) If any polynomial X, say of the fourth degree, is divided by any binomial of the form of $x - a$, the quotient will be a polynomial a unit lower in degree ; and the coefficients of the quotient can be expressed in terms of the coefficients of the proposed polynomial.

Demonstration.—Let $X = (x - a)\, Q + R.$

Now as R does not contain x, and as the divisor $x - a$ is of the first degree, therefore the quotient Q must be a unit lower in degree than the dividend X.

The form of the quotient will be as follows :

$$Q = A'_3\, x^3 + A'_2\, x^2 + A'\, x + A'_0$$

It is required to be proved that the values of A'_3, A'_2, A', A'_0 can be expressed in terms of the coefficients of the proposed equation.

Let $A_4 x^4 + A_3 x^3 + A_2 x^2 + A.x + A_0 =$ ˊ

$$(x-a)(A'_3 x^3 + A'_2 x^2 + A'.x + A'_0) + R =$$

$$A'_3 x^4 + (A'_2 - a A'_3) x^3 + (A' - a A'_2) x^2 + (A'_0 - a A') x - a A'_0 + R$$

As this is identical to the first member, the coefficients of like powers of x are equal to each other; and hence we have

$$A'_3 = A_4$$

$$A'_2 - a A'_3 = A_3 \quad \therefore \quad A'_2 = A_3 + a A'_3$$

$$A' - a A'_2 = A_2 \quad \therefore \quad A' = A_2 + a A'_2$$

$$A'_0 - a A' = A \quad \therefore \quad A'_0 = A + a A'$$

Also

$$R - a A'_0 = A_0 \quad \therefore \quad R = A_0 + a A'_0$$

Substitute for the values of the accented symbols, and we obtain

$$A'_3 = A_4$$
$$A'_2 = A_3 + a A_4$$
$$A' = A_2 + a A_3 + a^2 A_4$$
$$A'_0 = A + a A_2 + a^2 A_3 + a^3 A_4$$
$$R = A_0 + a A + a^2 A_2 + a^3 A_3 + a^4 A_4$$

The right hand members of the first four equations are the coefficients of the quotient Q, expressed in terms of the coefficients of the proposed equation; and the right hand member of the last equation is the value of the remainder after the division is executed. The coefficients when connected with x will give the following equation.

$$Q = A_4 x^3 + (A_3 + a A_4) x^2 + (A_2 + a A_3 + a^2 A_4) x + A + a A_2 + a^2 A_3 + a^3 A_4$$

It will be perceived that these coefficients are formed by a very simple law:

> The first is equal to the first of the proposed equation;
> the second is equal to the first multiplied by a, $+ A_3$;
> the third is equal to the second multiplied by a, $+ A_2$;
> the fourth is equal to the third multiplied by a, $+ A$;
> and the remainder is equal to the fourth multiplied by a, $+ A_0$.

(15.) A horizontal arrangement is the most convenient in practice. A few examples will be given for exercise.

EXAMPLES.

1. Required the quotient and remainder resulting from the division of

$$3\,x^4 + 8\,x^3 - 11\,x^2 + 6\,x - 19$$

by $x - 2$, where $a = 2$

	A_4	A_3	A_2	A	A_0	
Coefficients	3	+ 8 -	11 +	6	− 19	(2
		6 +	28 +	34	+ 80	
	3	14 +	17 +	40	+ 61	

Hence the quotient is

$$3x^3 + 14x^2 + 17x + 40$$

and the remainder is $+ 61$

2. Required, the quotient and remainder resulting from the division of

$$2\,x^3 - 29\,x^2 + x - 7$$

by $x - 7$, where $a = 7$

2	− 29 +	1	− 7	(7
	14 −	105	− 728	
	− 15 −	104	− 735	

The quotient is

$$2x^2 - 15x - 104$$

and the remainder is $- 735$

3. Required, the quotient and remainder resulting from the division of

$$x^6 - x^4 + x^3 - 13\,x - 987$$

by $x - 4$

1 + 0	− 1 +	1 +	0	− 13	− 987	(4
4	+ 16 +	60 +	244	+ 976	+ 3852	
4	+ 15 +	61 +	244	+ 963	+ 2865	

The quotient is

$$x^5 + 4\,x^4 + 15\,x^3 + 61\,x^2 + 244\,x + 963$$

and the remainder is $+ 2865$

By this example we see that the vacant coefficients must be supplied before proceeding to divide.

4. Required, the quotient and remainder arising from the division of

$$5\,x^4 - 17\,x^3 - 13\,x - 10004$$

by $x + 1$, where $a = -1$

$$
\begin{array}{rrrrrr}
5 & -\ 17 & +\ 0 & -\ 13 & -\ 10004 & (-1 \\
  & -\ 5 & +\ 22 & -\ 22 & +\ \ 35 & \\
\hline
  & -\ 22 & +\ 22 & -\ 35 & -\ \ 9969 & \\
\end{array}
$$

The quotient is

$$5\,x^3 - 22\,x^2 + 22\,x - 35$$

and the remainder is $- 9969$

5. Required, the quotient and remainder arising from the division of

$$3\,x^5 + 18\,x^4 - 60\,x^2 - 360\,x - 1$$

by $x + 6$, where $a = -6$

$$
\begin{array}{rrrrrr}
3 & +\ 18 & +\ 0 & -\ 60 & -\ 360 & -\ 1 \quad (-6 \\
  & -\ 18 & +\ 0 & +\ 0 & +\ 360 & +\ 0 \\
\hline
  & 0 & +\ 0 & -\ 60 & +\ 0 & -\ 1 \\
\end{array}
$$

The quotient is

$$3\,x^4 - 60\,x$$

and the remainder is $- 1$

6. Required, the quotient and remainder arising from the division of

$$\cdot 2\,x^4 - 1 \cdot 07\,x^3 + \cdot 319\,x^2 - 11 \cdot 2\,x + 1$$

by $x - \cdot 03$

$$
\begin{array}{l}
\cdot 2. - 1 \cdot 07 \;\; + \cdot 319 \;\;\;\; - 11 \cdot 2 \;\;\;\;\;\;\;\;\;\;\; + 1 \qquad\qquad (\cdot 03 \\
 \cdot 006 - \cdot 03192 + \;\;\;\; \cdot 0086124 - 0 \cdot 335741628 \\
\hline
- 1 \cdot 064 + \cdot 28708 - 11 \cdot 1913876 + \;\;\; \cdot.664258372
\end{array}
$$

Quotient

$$= \cdot 2\,x^3 - 1 \cdot 064\,x^2 + \cdot 28708\,x - 11 \cdot 1913876$$

the remainder is $+ \cdot 664258372$

PROPOSITION VI.

(16.) If a cubic equation has two imaginary roots, and its coefficients are real, the product of the two imaginary roots is positive.

Demonstration.—If the coefficients are real, the two imaginary roots must have either the form of $a + a_1 \sqrt{-1}, a - a_1 \sqrt{-1}$, or the form of $- a + a_1 \sqrt{-1}, - a - a_1 \sqrt{-1}$; but the product of either of these pairs is positive; hence the proposition is proved.

PROPOSITION VII. PROBLEM.

(17.) To determine the composition of the coefficients of a cubic or biquadratic equation.

Solution.—In proposition I. it has been proved, that an equation of the fourth degree, consists of four factors of the form of $(x - a_1)$ $(x - a_2)$ $(x - a_3)$ $(x - a_4)$, each of which is equal to zero: and in the corollary of the same proposition, it is shown that, by successive division, the biquadratic equation can be depressed to a cubic equation; the cubic to a quadratic equation; the quadratic to an equation of the first degree. Now the reverse process must necessarily elevate an equation from one degree to another; that is, the product of two of the factors will be a quadratic equation; the product of three factors, a cubic equation; and the product of four factors, a biquadratic equation; and so on; as it is evident the property is general.

Let these multiplications be executed, and we shall have

$$(x-a_1)(x-a_2) = x^2 -(a_1+a_2)x+a_1a_2 = 0$$

$$(x-a_1)(x-a_2)(x-a_3) =$$
$$x^3 -(a_1 + a_2 + a_3)x^2 + (a_1a_2 + a_1a_3 + a_2a_3)x - a_1a_2a_3 = 0$$

$$(x-a_1)(x-a_2)(x-a_3)(x-a_4)=$$
$$x^4 - (a_1+a_2+a_3+a_4)x^3 +(a_1a_2 + a_1a_3+a_1a_4+a_2a_3+a_2a_4+a_3a_4)x^2$$
$$-(a_1a_2a_3+a_1a_2a_4+a_1a_3a_4+a_2a_3a_4)x+a_1a_2a_3a_4 = 0$$

The law of the formation of the coefficients in these three equations becomes at once very evident.

In the quadratic equation, the second coefficient is the sum of the two roots with their signs changed ; and the final term is the product of the two roots.

In the cubic equation, the second coefficient is the sum of the three roots with their signs changed ; the third coefficient is the sum of their products two and two ; the final term is the product of the three roots with their signs changed.

In the biquadratic equation, the second coefficient is the sum of the four roots with their signs changed ; the third coefficient is the sum of their products two and two : the fourth, is the sum of their products three and three with their signs changed ; the final term is the product of the four roots.

COROLLARY 1.

(18.) If the coefficient of the second term of either of these three equations be 0, the sum of the positive roots must equal the sum of the negative roots.

This is evident from the law of the formation of the second coefficient.

COROLLARY 2.

(19.) The final term of the cubic equation is divisible by any one of its roots ; or by the product of any two of its roots.

The final term of the biquadratic equation is divisible by any one of its roots, or by the product of any two of its roots ; or by the product of any three of its roots.

This is also evident from the law of the formation of the absolute term

PROPOSITION VIII.

(20.) If the coefficients of a cubic equation are real, it has at least one real root, of which the sign is contrary to that of the final term of the equation.

Demonstration.—If the three roots are positive, then by the law of formation of the final term (prop. vii.) it will be negative; if the three roots are negative, by the same law the final term will be positive; if two roots are imaginary, their product is positive (prop. vi.); if a positive root is multiplied into this positive product, and the sign changed (prop. vii.), the final term will be negative; if a negative root is multiplied into the positive product of the two imaginary roots, and the sign changed, the final term will be positive : therefore the proposition is demonstrated.

PROPOSITION IX.

(21.) If the roots of a cubic or biquadratic equation are known to be real, and the signs of the equation are alternately positive and negative, the roots are all positive.

Demonstration.—Let $x^2 - (a_1 + a_2) x + a_1 a_2 = 0$ be a general quadratic equation, formed by the multiplication of the two factors $(x - a_1) (x - a_2)$, a_1 and a_2 being positive roots ; now if this quadratic is multiplied by a third factor $x - a_3$, a_3 being a positive root, the product will be a cubic equation with signs alternately positive and negative, as represented in proposition vii. If this cubic equation is multiplied by a fourth factor x—a_4 , a_4 being a positive root, the result will be a biquadratic equation with signs alternately positive and negative, as shown in the same proposition : now there is no other combination of real roots which can produce this form of signs : therefore, if the roots are known to be real, this alternate form of signs proves them to be all positive.

PROPOSITION X.

(22.) If the roots of a cubic equation whose second term is absent are known to be real, two of its roots will have the same sign as the final term, and the remaining root will be of the opposite sign.

Demonstration.—1. If two roots are positive the remaining root must be negative; for by proposition vii., corollary 1, the sum of the positive roots must be equal to the sum of the negative roots; but if two roots are positive and one negative, their continued product with changed sign will be positive; therefore, (prop. vii.) the final term will be positive.

2. If two roots are negative and one positive, for the same reasons just given, the final term will be negative ; therefore, if the roots are known to be real, two of the roots must have the same sign as the final term, and the remaining root must be of the opposite sign.

CHAPTER II.

(23.) Equations of all orders are susceptible of a great number of transformations : but only such as have an immediate bearing upon the numerical solution of equations of the third and fourth degrees, will, in the present chapter, receive much attention.

To transform an equation, is to change it into another equation, whose roots have a certain given relation to the roots of the proposed equation : this can be effected, although the roots of the proposed and transformed equations are unknown. It is often the case, that the roots of the transformed equation can be much more easily found, than those of the proposed equation ; and when the former become known, the roots of the latter can be immediately derived from them. Moreover, transformed equations often discover to us the number of real and imaginary roots existing in a proposed equation, which will be found an object of great importance in practical solution. But one of the principal uses of transformation is the finding of the first figure of a root; and thus, in a very simple manner, solving a problem which has occupied the attention of mathematicians for several centuries, and upon which volumes have been written. But the utility of transformation will become much more apparent as we proceed, as exhibited in connection with the subjects to be elucidated.

PROPOSITION I.

(24.) If the alternate signs of the coefficients of a complete cubic or biquadratic equation are changed, commencing with the second, the roots will be the same as those of the proposed equation, but with changed signs.

Demonstration.—Let $x^4 + A_3 x^3 + A_2 x^2 + A x + A_0 = 0$ be the proposed equation; assume $x_1 = -x$; now when x has any particular value, it is evident that $-x_1$ will have the same numerical value but with a contrary sign; thus $x = -x_1$; substitute $-x_1$ in the proposed equation, and we shall have

$$(-x_1)^4 + A_3 (-x_1)^3 + A_2 (-x_1)^2 + A (-x_1) + A_0 = 0$$

that is

$$x^4 - A_3 x^3 + A_2 x^2 - Ax + A_0 = 0$$

and if $-x_1$ is substituted in a cubic equation we shall obtain

$$x^3 - A_2 x^2 + Ax - A_0 = 0$$

Thus it is proved that by changing the alternate signs of these equations, commencing with the second term, the signs of all the roots are changed, but not their numerical values.

Let it be remembered that the equation must be made *complete*, if lacking any terms, before the alternate signs are changed; for example : let it be required to transform the equation

$$x^4 + 3x^2 - 5x + 7 = 0$$

into another whose roots shall be numerically the same, but with contrary signs. The proposed equation, with the vacant term supplied, will be thus

$$x^4 + 0 x^3 + 3 x^2 - 5 x + 7 = 0$$

and the required transformed equation will be

$$x_1^4 + 3 x_1^2 + 5 x_1 + 7 = 0$$

We will add another example. Transform the cubic equation

$$x^3 + 19 x + 16 = 0$$

into another whose roots shall be numerically the same, but with changed signs.

The proposed equation, with the vacant term supplied, is as follows :

$$x^3 + 0\,x^2 + 19\,x + 16 = 0$$

The transformed equation is

$$x_1^3 + 19\,x_1 - 16 = 0$$

COROLLARY.

(25.) If the roots of a cubic or biquadratic equation are known to be real, and the terms of the equation are all positive, all the roots will be negative.

Demonstration.—In proposition IX, chap. I, it was proved, that when the roots are real, and the alternate terms of the equation positive and negative, the roots are all positive : and in this proposition it is proved that the signs of the roots will be changed by changing the signs of the alternate terms, beginning with the second term : but to change the signs of the alternate terms will render all the terms of such equation positive ; therefore, if the roots are real, and all the terms positive, all the roots will be negative.

PROPOSITION II.   PROBLEM.

(26.) To transform an equation, say of the fourth degree, into another whose roots shall be equal to those of the proposed equation multiplied by a given quantity.

Let $x^4 + A_3\,x^3 + A_2\,x^2 + A\,x + A_0 = 0$ be the proposed equation. It has been proved in art. (17.) that A_3 is equal to the sum of the roots with changed signs, that A_2 is equal to the sum of their products two and two, that A is equal to the sum of their products three and three, and that A_0 is equal to the product of the four roots.   Now it is required to multiply each root by a given quantity m ; execute the multiplication in each coefficient ; thus

$$ma_1 + ma_2 + \&c. = m\,A_3$$
$$ma_1 \times ma_2 + ma_1 \times ma_3 + \&c. = m^2 A_2$$
$$ma_1 \times ma_2 \times ma_3 + ma_1 \times ma_2 \times ma_4 + \&c. = m^3 A$$
$$ma_1 \times ma_2 \times ma_3 \times ma_4 = m^4 A_0$$

The transformed equation will therefore become

$$x_1^4 + m\,A_3\,x_1^3 + m^2\,A_2\,x_1^2 + m^3\,A\,x_1 + m^4\,A_0 = 0$$

m may be any quantity, positive or negative, fractional or integral. It is also evident that this law is general, and applicable to equations of all degrees.

As an example, let it be required to transform the equation

$$x^3 - 11\,x^2 + 23x + 35 = 0$$

whose roots are -1, 5, and 7, into another whose roots shall be twice as great.

Multiply the coefficients by 2, 4, and 8, and the transformed will be

$$x^3 - 22\,x^2 + 92\,x + 280 = 0$$

the roots of which are -2, 10, and 14, as may be proved by substituting each of these numbers for x.

Again, transform the equation

$$x^3 - 63x + 162 = 0$$

whose roots are 3, 6, and -9, into another whose roots shall be only one-third as great.

Supply the vacant term

$$x^3 + 0\,x^2 - 63x + 162 = 0$$

Multiply the second, third, and fourth terms respectively by $\frac{1}{3}$, $\frac{1}{9}$, and $\frac{1}{27}$, and we shall obtain the transformed equation,

$$x^3 - 7x + 6 = 0$$

whose roots are 1, 2, and -3, being only one-third of those of the proposed.

(27.) It is often desirable to reduce the coefficient of the leading term to unity; but when this is done by common division, it frequently renders a part or all of the other coefficients fractional: to avoid this, multiply the roots by a quantity m equal to the first coefficient, and then divide by this coefficient: or, which amounts to the same thing, multiply the third term by m, the fourth by m^2, the fifth by m^3, and so on, and merely remove the coefficient of the first term : the result will give an equation whose roots are m times as great and whose coefficients are integral.

(28.) By this method all fractional coefficients can be easily removed : first, find a common multiple of all the denominators ; and secondly, multiply the roots of the equation by this common multiple. For example, transform the equation

$$x^3 - \tfrac{1}{2} x^2 + \tfrac{1}{8} x - \tfrac{1}{6} = 0$$

into another whose coefficients shall be integral.

In this case, the common multiple is 6 ; hence multiply the roots by 6, and the result will be

$$x^3 - 3 x^2 + 12 x - 36 = 0$$

(29.) Decimals may be removed from the coefficients without altering the value of the roots, by simply multiplying each term of the equation by 10, or 100, or 1000, &c. ; that is, the multiplier must contain the same number of cyphers as the highest number of decimals in any coefficient : as, for instance, remove the decimals from the equation

$$\cdot012 x^4 + 13\cdot5 x^3 - 1\cdot4 x^2 + \cdot21 x - 6 = 0$$

Multiply each term by 1000, and the result will be

$$12 x^4 + 13500 x^3 - 1400 x^2 + 210 x - 6000 = 0$$

This is the same as removing the decimal point three figures to the right, which is equivalent to expunging it. The roots of this equation are the same as in the proposed. But if it be required to reduce the coefficient 12 to unity without introducing fractions, then the roots must be multiplied by 12 ; that is, the third, fourth, and fifth terms must be multiplied respectively by 12, $(12)^2$, and $(12)^3$ (see art. 27), expunging the leading coefficient.

PROPOSITION III.   PROBLEM.

(30.) To transform an equation into another the roots of which shall be greater or less than those of the proposed equation by any given quantity.

The method about to be explained is general; but it may be more simply illustrated by selecting an equation of some specified degree; say, for instance, that of the fourth degree, namely,

$$A_4 x^4 + A_3 x^3 + A_2 x^2 + A x + A_0 = 0$$

Let this be the proposed equation; and let it be required to transform it into another equation of the same degree, but whose roots shall be greater or less than those of the proposed by some given quantity r. Let the roots of the transformed equation be represented by x', it is evident that

$$x' + r = x$$

consequently
$$x' = x - r$$

By this last equation it will be seen that when r is positive, x' is less than x by the quantity r; but when r is negative, x' will be greater than x: the minus sign before r refers only to its numerical coefficient or unity.

Let the transformed equation, by substituting $x' + r$ for x, be

$$A'_4 x'^4 + A'_3 x'^3 + A'_2 x'^2 + A' x' + A'_0 = 0$$

Substitute in this $x - r$ for x', and we shall have

$$A_4 (x - r)^4 + A'_3 (x - r)^3 + A'_2 (x - r)^2 + A' (x - r) + A'_0 =$$
$$A_4 x^4 + A_3 x^3 + A_2 x^2 + A x + A_0$$

All the coefficients of this second member of the equation are known; but those of the first member, with the exception of A_4, are unknown. Now it is evident that if the first member be divided by $x - r$, the remainder will be A'_0: but as the two members are identical, the same remainder must result from dividing the second member by $x - r$: let this division be executed; the remainder will be A'_0, and the quotient will be

$$A_4 (x-r)^3 + A'_3 (x-r)^2 + A'_2 (x-r) + A'$$

Also dividing this by $x - r$, we obtain for remainder A', and for the quotient
$$A_4 (x-r)^2 + A'_3 (x-r) + A'_2$$

Again dividing this latter by $x - r$, the resulting remainder is A'_2, and the quotient

$$A_4 (x-r) + A'_3$$

And finally dividing this by $x - r$, we have for the last remainder A'_3, and for the final quotient A_4.

Thus all the coefficients of the transformed equation, namely, A_4, A'_3, A'_2, A', A'_0 become known; being obtained by the simple process of successive division by $x - r$.

An easy method of performing these divisions is given in articles (14) and (15).

For this superior and expeditious method of finding the coefficients of the transformed equation, the author is indebted to an excellent work on the Cubic Equation by Prof. J. R. YOUNG.

EXAMPLES.

1. Transform the equation

$$x^3 - 8x^2 + 5x - 20 = 0$$

into one whose roots shall be the roots of this increased by 3.

A_3	A_2	A	A_0
1	− 3	+ 5	− 20	(− 8 = r, r being negative
	− 3	+ 18	− 69	
	− 6	+ 23	− 89	$\therefore A'_0 = - 89$
	− 3	+ 27		
	− 9	+ 50		$\therefore A' = 50$
	− 3			
	−12			$\therefore A'_2 = - 12$

Hence the transformed equation is $x'^3 - 12x'^2 + 50x' - 89 = 0$

2. Transform the equation

$$x^3 - 4x^2 - x + 3 = 0$$

into another whose roots shall be less than the roots of this by 2.

$$1 - 4 - 1 \quad + \quad 3 \; (2 = r, \; r \text{ being positive.}$$
$$\quad 2 - 4 \quad - 10$$
$$\overline{\quad -2 - 5 \quad -7} \quad A_0 = 7$$
$$\quad 2 \quad 0$$
$$\overline{\quad 0 \; -5} \qquad A_1 = -5$$
$$\quad 2$$
$$\overline{x'^3 + 2x'^2 - 5x' - 7 = 0}$$

3. Transform the equation

$$x^4 + x^3 - x^2 + x - 10 = 0$$

into another whose roots shall be the roots of this diminished by 1.

$$1 + 1 - 1 + 1 - 10 \; (1$$
$$\quad 1 + 2 + 1 + \; 2$$
$$\overline{\quad 2 + 1 + 2}$$
$$\quad 1 + 3 + 4$$
$$\overline{\quad 3 + 4}$$
$$\quad 1 + 4$$
$$\overline{\quad 4}$$
$$\quad 1$$
$$\overline{x'^4 + 5\,x'^3 + 8\,x'^2 + 6\,x' - 8 = 0}$$

4. Transform the equation

$$2\,x^4 - 16\,x^3 + 7\,x^2 + 10\,x - 100 = 0$$

into one whose roots shall be the roots of this diminished by 2.

$$2 - 16 + \; 7 + 10 - 100 \; (2$$
$$\quad 4 - 24 - 34 - \quad 48$$
$$\overline{\quad -12 - 17 - 24}$$
$$\quad 4 - 16 - 66$$
$$\overline{\quad -8 - 33}$$
$$\quad 4 - \; 8$$
$$\overline{\quad -4}$$
$$\quad 4$$
$$\overline{2\,x'^4 + 0\,x'^3 - 41\,x'^2 - 90\,x' - 148 = 0}$$

5. Transform the equation

$$2 x^4 - 8 x^3 + 12 x^2 - 8 x + 3 = 0$$

into another whose roots shall be the roots of this diminished by 1.

$$
\begin{array}{l}
2 - 8 + 12 - 8 + 3 \,(1 \\
\quad\ 2 - \ 6 + 6 - 2 \\
\overline{-6 + \ 6 - 2} \\
\quad\ 2 - \ 4 + 2 \\
\overline{-4 + \ 2} \\
\quad\ 2 - \ 2\ . \\
\overline{-2} \\
\quad\ 2 \\
\hline
2\,x'^4 + 0\,x'^3 + 0\,x'^2 + 0\,x' + 1 = 0
\end{array}
$$

that is $\qquad 2\,x'^4 + 1 = 0$

6. Transform the equation

$$X = x^3 - 12 v^2 + x - 1 = 0$$

into another whose roots shall be the roots of this diminished by 20·

$$
\begin{array}{l}
1 - 12 \ + \quad 1 \qquad - 1 \ (20 \\
\quad\ 20 + 160 \ + 3220 \\
\overline{\quad\ 8 + 161} \quad\ \cdot \\
\quad\ 20 + 560 \\
\overline{\quad 28} \\
\quad\ 20 \\
\hline
X_1 = x'^3 + 48 v'^2 + 721 v' + 3219 = 0
\end{array}
$$

7. Transform $X_1 = 0$, or the resulting equation just obtained, into an equation whose roots shall be the roots of this diminished by 1.

$$
\begin{array}{l}
1 \ + 48 \ + \ 721 \ + 3219 \ (1 \\
\quad\ 1 + \ 49 \ + \ 770 \\
\overline{\quad 49 + 770} \\
\quad\ 1 + \ 50 \\
\overline{\quad 50} \\
\quad\ 1 \\
\hline
X_{11} = x''^3 + 51 a''^2 + 820 v'' + 3989 = 0
\end{array}
$$

8. Transform $X_{\shortmid\shortmid} = 0$ into an equation whose roots shall be less than this by ·3.

$$
\begin{array}{rllll}
1 & +\ 51 & +\ 820 & +\ 3989 & (\cdot 3 \\
  & \cdot 3 & +\ 15\cdot 39 & +\ 250\cdot 617 & \\
\cline{2-3}
  & 51\cdot 3 & +\ 835\cdot 39 & & \\
  & \cdot 3 & +\ 15\cdot 48 & & \\
\cline{2-3}
  & 51\cdot 6 & & & \\
  & 3 & & & \\
\end{array}
$$

$$X_{\shortmid\shortmid\shortmid} = x'''^{3} + 51\cdot9x'''^{2} + 850\cdot87x''' + 4239\cdot617 = 0$$

The roots of $X_{\shortmid\shortmid\shortmid} = 0$ are equal to the roots of $X = 0$ (Ex. 7) diminished by 21·3.

The student should make himself familiar with this class of transformations, as he will find the method of great utility in the numerical solution of equations of all orders, and especially in the solution of equations of a higher degree than the biquadratic.

PROPOSITION IV. PROBLEM.

(31.) To transform an equation into another whose second term shall be absent.

If the equation is of the nth degree, reduce the coefficient of the leading term to unity, art. (27.), and then proceed to diminish the roots by minus the nth part of the second coefficient, according to the method given in proposition III.

EXAMPLES.

1. Transform the equation

$$x^3 - 15x^2 + 4x - 7 = 0$$

into another whose second term shall be absent.

$$
\begin{array}{rrrrl}
\cdot 1 & -\ 15 & +\ 4 & -\ 7 & (5 \\
       & 5 & -\ 50 & -\ 230 & \\
\cline{2-3}
       & -\ 10 & -\ 46 & & \\
       & 5 & -\ 25 & & \\
\cline{2-3}
       & -\ 5 & & & \\
       & 5 & & & \\
\end{array}
$$

Transformed equation $x'^3 + 0x'^2 - 71x' - 237 = 0$

In this example the equation is of the third degree; therefore $n = 3$. Hence *minus* the third part of -15, the coefficient of the second term, is equal to $+5$, which is the amount by which the roots are to be diminished.

2. Transform the equation

$$x^3 + 6x^2 - 2x + 1 = 0$$

into another in which the second term shall be absent.

```
1  + 6  -  2  +  1  (- 2
   - 2  -  8  + 20
     4  - 10
   - 2  -  4
   ----
     2
   - 2
```

Transformed equation $\quad x'^3 + 0.x'^2 - 14.x' + 21 = 0$

3. Transform the equation

$$x^4 - 12x^3 + x^2 + 17x - 84 = 0$$

into another whose second term shall be absent.

```
1  - 12  +  1  +  17  -  84  (3
      3  - 27  -  78  - 183
   -  9  - 26  -  61
      3  - 18  - 132
   -  6  - 44
      3  -  9
   -  3
      3
```

Transformed equation $x'^4 + 0.x'^3 - 53.x'^2 - 198.x' - 267 = 0$

If the coefficient of the second term is not divisible by n, the highest power of the leading quantity, and if it is desirable not to introduce fractional coefficients, multiply the roots by n, art. (26.), after which divide the second coefficient by minus n, and proceed to diminish as above. The roots of the second transformed equation will be those of the first transformed equation diminished by the minus nth part of its second coefficient; and the roots of the first transformed equation will be those of the original equation multiplied by n.

PROPOSITION V. PROBLEM.

32. *Required to transform the equation*

$$x^3 + Ax + A_0 = 0$$

into another, the roots of which are three of the differences of the roots of the proposed equation.

Let a_1, a_2, a_3, denote the roots of the proposed equation; then, by articles (17.) and (18.),

$$a_1 + a_2 + a_3 = 0$$
$$a_1a_2 + a_1a_3 + a_2a_3 = A$$
$$a_1a_2a_3 = -A_0$$

Squaring the first of these equations, we have

$$a_1^2 + a_2^2 + a_3^2 = -2A$$

Now there are six differences between the roots of all cubic equations; consequently the roots of the required transformed equation must be three out of the six differences. Let the six differences be as follows :—

1st . . . $a_2 - a_1$	4th . . , $a_1 - a_2$
2nd . . $a_1 - a_3$	5th . . . $a_3 - a_1$
3rd . . $a_3 - a_2$	6th . . . $a_2 - a_3$

Either the first or the last three may be selected; the result will be the same, but with contrary signs. Let the first three, for instance, denote the roots of the required transformed equation; then we shall have for the sum of the roots

$$(a_2 - a_1) + (a_1 - a_3) + (a_3 - a_2) = 0$$

for their products two by two

$$(a_2 - a_1)(a_1 - a_3) + (a_2 - a_1)(a_3 - a_2) + (a_1 - a_3)(a_3 - a_2) =$$
$$-(a_1^2 + a_2^2 + a_3^2) + a_1a_2 + a_1a_3 + a_2a_3 = 3A$$

Thus all but the final term of the transformed equation of differences becomes known. To obtain the absolute term, we must find the equation of the squares of the differences, from which the final term of the required equation is easily derived.

The three roots or squares of the differences will be

$$(a_2 - a_1)^2, \ (a_1 - a_3)^2, \ (a_3 - a_2)^2$$

now

$$(a_2 - a_1)^2 = a_2^2 - 2\,a_2 a_1 + a_1^2 = a_2^2 + a_1^2 + a_3^2 - 2\,a_2 a_1 - a_3^2 =$$

$$a_2^2 + a_1^2 + a_3^2 - \frac{2 a_2 a_1 a_3}{a_3} - a_3^2 = -2\,A + \frac{2\,A_0}{a_3} - a_3^2 = y^2$$

y^2 being one of the roots. Substitute in this last equation x for a_3 and we have

$$y^2 = -2A + \frac{2\,A_0}{x} - x^2$$

that is

$$x^3 + (2\,A + y^2)\,x - 2\,A_0 = 0$$

the proposed is

$$x^3 + Ax + A_0 = 0$$

From these two equations eliminate x by subtracting, and we obtain

$$x = \frac{3\,A_0}{A + y^2}$$

Substitute this value of x in the proposed equation, and reduce; the result will be

$$y^6 + 6\,A\,y^4 + 9\,A^2\,y^2 + 4\,A^3 + 27\,A_0^2 = 0$$

This is an equation of the squares of the differences: though of the sixth degree, yet there are only even powers of the unknown quantity involved: y^2 having three values, namely, the squares of the three differences. Therefore by making $y^2 = y$, and substituting, the equation will become of the third degree.

Now the absolute term is equal to the product of the squares of the differences with their signs changed, (art. 17.) Hence we have

$$-(a_2 - a_1)^2\,(a_1 - a_3)^2\,(a_3 - a_2)^2 = \quad 4\,A^3 + 27\,A_0^2$$

and

$$(a_2 - a_1)^2\,(a_1 - a_3)^2\,(a_3 - a_2)^2 = -4\,A^3 - 27\,A_0^2$$

therefore

$$(a_2 - a_1)\,(a_1 - a_3)\,(a_3 - a_2) = \pm\,\sqrt{-4\,A^3 - 27\,A_0^2}$$

Thus we have found the final term of the equation of differences. And we have proved above that the sum of the terms is equal to 0 ; and that the sum of the products two by two is equal to 3 A ; therefore by substituting these coefficients the equation of differences will stand as follows

$$y^3 + 3 A y \pm \sqrt{- 4 A^3 - 27 A_0^2} = 0$$

When A is positive, both terms under the radical sign become negative ; hence the final term is imaginary.

When A is negative, $- 4 A^3$ becomes positive ; but if less than $- 27 A_0^2$, the final term is still imaginary : but when A is negative, and $- 27 A_0^2$ is less than $- 4 A^3$, the final term is real.

(33.) This equation, discovered by the author [*] about five years ago, is one of the highest importance in simplifying the numerical solution of the cubic and biquadratic equations. If the equation has been discovered by others, the author is not aware of it ; certain it is, that no mention of any such discovery is to be found in the mathematical works which have come under his observation.

The great value of this equation in detecting imaginary roots, in finding the first figure of a root, and in numerous other inquiries, will become abundantly manifest at a future stage of the work.

PROPOSITION VI.. PROBLEM.

(34.) *Required to transform the equation*

$$X = x^3 + Ax + A_0 = 0$$

into another the roots of which are the second differences of the roots of the proposed equation.

In proposition v. we found an equation of first differences, namely,

$$Y = y^3 + 3 A y \pm \sqrt{- 4 A^3 - 27 A_0^2} = 0$$

It is evident, if we transform the equation $Y = 0$ into another, say $Z = 0$, whose roots shall be the differences between the roots

<hr />

[*] The *Equation of Differences* was also discovered about the same time by my son, Orson Pratt jun., of Utah Territory, U.S.A.; each being uninformed in regard to the other's discovery, as I was at the time in New York city, about three thousand miles from him. We exchanged letters upon the discovery of nearly the same date.

of $Y = 0$, that the roots of $Z = 0$ will be the second differences of the roots of $X = 0$.

Let $Y = 0$ receive the specified transformation; thus

$$Z = z^3 + 9\,A\,z \pm \sqrt{-4\,(3\,A)^3 - 27\,(\sqrt{-4\,A^3 - 27\,A_0^2})^2} = 0$$

that is $\qquad z^3 + 9\,A\,z \pm 27\,A_0 = 0$

This is an equation of second differences between the roots of $X = 0$: but if this be compared with the proposed, it will at once be seen, that the roots of the equation of second differences are the roots of the proposed equation multiplied by three.

PROPOSITION VII.   PROBLEM.

(35.) *Required to transform the equation*

$$Y = y^3 + A\,y + A_0 = 0$$

into another in which the differences of the roots shall be equal to the roots of $Y = 0$.

It is evident that this is the reverse of proposition v.

Let the required transformed equation be

$$X = x^3 + A'x + A'_0 = 0$$

Transform this into the equation of differences, and we have

$$X' = x'^3 + 3\,A'\,x' \pm \sqrt{-4\,A'^3 - 27\,A_0'^2} = 0$$

Now we have the two identical equations $Y = 0$ and $X' = 0$; therefore, by equating the coefficients, we obtain

$$A = 3\,A'\,\cdot$$

and

$$A_0 = \pm \sqrt{-4\,A'^3 - 27\,A_0'^2}$$

therefore $\qquad A' = \dfrac{A}{3}$

substituting $\qquad A_0 = \pm \sqrt{-4\left(\dfrac{A}{3}\right)^3 - 27\,A'_0{}^2}$

hence $\qquad A'_0 = \pm \dfrac{1}{27}\sqrt{-4\,A^3 - 27\,A_0^2}$

therefore $\qquad X = x^3 + \dfrac{A}{3}\,x \pm \dfrac{1}{27}\sqrt{-4\,A^3 - 27\,A_0^2} = 0$

This is the required transformed equation. This equation will also be found very useful in obtaining the two remaining roots of an equation after one is known, as will be illustrated hereafter.

CHAPTER III.

(36.) In the future part of this treatise, we shall, as a general thing, drop from the equations Y and Z the use of the capital letters A_2, A, A_0, &c. as not being quite so convenient as smaller ones. And in their place we shall adopt the letters b and c, and b_1 and c_1: thus $Y = y^3 + by + c = 0$: and for its equation of differences $Z = z^3 + b_1 z + c_1 = 0$.

PROPOSITION I.

(37.) A cubic equation, whose second term is wanting, cannot have one imaginary and two real roots.

Demonstration.—The roots of every equation, wanting the second term, are so related that each root is the sum of the other two with their signs changed, articles (17.) and (18.); but the sum of two real roots is real; therefore, such an equation cannot have one imaginary and two real roots.

PROPOSITION II.

(38.) If a cubic equation, lacking the second term, contains imaginary roots, and b and c are real, and either positive or negative, the equation can have only two imaginary roots.

Demonstration.—Let p be the product of any two of the roots; let s be the remaining root : then

$$b = p - s^2$$

and
$$c = ps$$

If either p or s is imaginary, and the other factor real, c will become imaginary; but by hypothesis c is real; therefore p and s must both be imaginary, or both be real: suppose both imaginary, and let

$$p = a + a_1 \sqrt{-1}$$
and $$s = a - a_1 \sqrt{-1}$$
then $c = $ $$ps = a^2 + a_1^2$$
and $b = p - s^2 = a + a_1 \sqrt{-1} - a^2 + 2aa_1 \sqrt{-1} + a_1^2$; hence,

b is imaginary; but by hypothesis b is real; therefore p and s cannot both be imaginary; hence, as proved above, they must both be real; therefore s, which is one of the roots, must be real: and as the equation cannot, by the last article, contain two real roots, the other two, whose product equals p, must be imaginary.

PROPOSITION III.

(39.) If a cubic equation whose second term is wanting contains imaginary roots, and b is real, and either positive or negative, and c imaginary, and either positive or negative, the equation must have three imaginary roots.

Demonstration.—Let $p = $ the product of any two of the roots; let $s = $ the remaining root:

hence
$$b = p - s^2$$
$$c = ps$$

If any one of the roots s is real, p must be imaginary, when c is imaginary; and if p is imaginary, b must also be imaginary; but by hypothesis b is real; therefore s cannot be real; therefore the equation must have three imaginary roots.

PROPOSITION IV.

(40.) If a cubic equation contains imaginary roots, its equation of differences will also contain imaginary roots; and *vice versa*, if the equation of differences contains imaginary roots, the proposed equation will contain imaginary roots.

Demonstration.—The number of differences between the three roots is six; three positive and three negative; but an equation of differences can be formed, consisting of only three of these differences as roots, art. (32.) If the roots of the proposed equation are real, any three of the differences will be real. If two or three roots of the proposed equation are imaginary, some of any three differences must be imaginary: for though the difference between two imaginary roots may be real, yet the differences between either of the two and the remaining root, which is the sum of the two with the signs changed, articles (17.) and (18.), must necessarily be imaginary.

And vice versa, if the equation of differences have two or three imaginary roots, the proposed equation will have imaginary roots; for the equation of second differences will, according to the above reasoning, have imaginary roots; but it has been proved, art. (34.), that the roots of an equation of second differences are three times the value of the roots of the proposed equation; and therefore, if any of the former are imaginary, some of the latter must likewise be imaginary.

CHAPTER IV

EQUAL ROOTS.

PROPOSITION I.

(41.) *When* $c = \pm \dfrac{2 \sqrt{-b^3}}{3 \sqrt{3}}$, *then*

$$Y_{,} = y^3 + by \pm \frac{2 \sqrt{-b^3}}{3 \sqrt{3}} = 0$$

will be a general equation for two equal roots.

Demonstration.—Transform $Y_{,}$ into Z ; thus

$$Z = z^3 + 3\,b\,z \pm \sqrt{-4\,b^3 - 27 \left(\frac{2 \sqrt{-b^3}}{3 \sqrt{3}} \right)^2} = 0$$

As the quantity under the radical sign reduces to 0, one of the roots of Z must be 0 ; therefore, the difference between two of the roots of $Y_{,}$ is equal to 0 ; therefore two roots must be equal ; therefore $Y_{,} = 0$ is a general equation for two equal roots.

COROLLARY 1.

(42.) If b, in equation $Y_{,} = 0$, is real and negative, and the absolute term either positive or negative, then equation $Y_{,} = 0$ will not only have two equal roots, but all three of its roots will be real.

For as the coefficients of both $Y_{,}$ and Z are real, the roots must be real.

The roots $Y_{,} = 0$ are $\pm \sqrt{-\dfrac{b}{3}}$, $\pm \sqrt{-\dfrac{b}{3}}$, $\mp 2 \sqrt{-\dfrac{b}{3}}$

The roots $Z = 0$ are 0 , $\pm \sqrt{-3b}$

(b being negative, the above values under the radical are positive quantities.)

COROLLARY 2.

(43.) If b, in equation $Y_{,} = 0$ is real and positive, and the absolute term either positive or negative, then the equation will have three imaginary roots, two of which will be equal.

As b is real and the absolute term of $Y_{,} = 0$ imaginary; therefore, by art. (39), $Y_{,} = 0$ has three imaginary roots.

The roots of $Y_{,} = 0$ are $\pm \sqrt{-\dfrac{b}{3}}$, $\pm \sqrt{-\dfrac{b}{3}}$, $\mp 2\sqrt{-\dfrac{b}{3}}$

The roots of $Z = 0$ are 0 , $\pm \sqrt{-3b}$

(By hypothesis b is positive, therefore $-\dfrac{b}{3}$ is negative.)

COROLLARY 3.

(44.) If b, in the general equation $Y_{,} = 0$, is imaginary, and either positive or negative, and the absolute term either positive or negative, then the equation will have three imaginary roots, two of which will be equal.

For the two equations $Y_{,}$ and Z will become

$$Y = y^3 + b\sqrt{-1}\, y \pm \frac{2\sqrt{-(b\sqrt{-1})^3}}{3\sqrt{3}}$$

$$Z = z^3 + 3b\sqrt{-1}\, z \pm \sqrt{-4(b\sqrt{-1})^3 - 27\left(\frac{2\sqrt{-(b\sqrt{-1})^3}}{3\sqrt{3}}\right)^2} = 0$$

The final term of $Z = 0$ is 0; therefore the equation $Y_{,} = 0$ has two equal roots; and the final term of $Y_{,} = 0$ being imaginary the equation has three imaginary roots.

The roots of $Y_{,} = 0$ are

$$\pm \sqrt{-\frac{b}{3}\sqrt{+\sqrt{-1}}},\ \pm \sqrt{-\frac{b}{3}\sqrt{+\sqrt{-1}}},\ \mp 2\sqrt{-\frac{b}{3}\sqrt{+\sqrt{-1}}}$$

The roots of $Z = 0$ are 0, $\pm \sqrt{-3\,b}\,\sqrt{+\sqrt{-1}}$

(45. The same relations exist, if b consists of two terms, one real and the other imaginary.

$$Y_{,} = y^3 + (a_{,} + b\,\sqrt{-1})\,y \pm \frac{2\sqrt{-(a_{,}+b\,\sqrt{-1})^3}}{3\,\sqrt{3}} = 0.$$

The roots of which are

$$\pm \sqrt{\frac{-(a_{,}+b\,\sqrt{-1})}{3}},$$

$$\pm \sqrt{\frac{-(a_{,}+b\,\sqrt{-1})}{3}},$$

$$\mp 2\sqrt{\frac{-(a_{,}+b\,\sqrt{-1})}{3}}$$

We have also

$$Z = z^3 + 3\,(a_{,}+b\,\sqrt{-1})\,z$$

$$\pm \sqrt{-4\,(a_{,}+b\,\sqrt{-1})^3 - 27\left(\frac{2\sqrt{-(a_{,}+b\sqrt{-1})^3}}{3\,\sqrt{3}}\right)^2} = 0;$$

the roots of this are 0, $\pm \sqrt{-3\,(a_{,}+b\,\sqrt{-1})}$

The same relations exist if b in the general equation $Y_{,}$ consists of any number of terms, real or imaginary, positive or negative, powers or roots. Indeed, it has already been demonstrated that the proposition is general; but the above corollaries and example may serve to impress its generality more forcibly upon the mind.

(46.) If b in the general equation $Y_{,}$ becomes 0, both y^3 and z^3 each become $= 0$; therefore, each have three equal roots equal to nothing.

PROPOSITION II.

(47.) *When* A_0, *in the general equation*

$$X = x^3 + A_2x^2 + Ax + A_0 = 0,$$

equals $\dfrac{9\,A_2\,A - 2\,A_2{}^3 \pm 2\,(-3\,A + A_2{}^2)^{\frac{3}{2}}}{27}$, *then will the complete cubic equation*

$$X_1 = x^3 + A_2x^2 + Ax + \dfrac{9\,A_2\,A - 2\,A_2{}^3 \pm 2\,(-3\,A + A_2{}^2)^{\frac{3}{2}}}{27} = 0$$

contain two equal roots.

Demonstration. Transform $X = 0$ into an equation whose second term is absent (31), and we shall have

$$Y = y^3 + (A - \tfrac{1}{3}A_2{}^2)\,y + \dfrac{2}{27}A_2{}^3 - \dfrac{A_2A}{3} + A_0 = 0$$

By prop. I. we have $c = \pm\,\dfrac{2\,\sqrt{-b^3}}{3\,\sqrt{3}} = \dfrac{2}{27}A_2{}^3 - \dfrac{A_2A}{3} + A_0$

Restore the value of $\sqrt{-b^3} = \sqrt{(-A + \tfrac{1}{3}A_2{}^2)^3}$ in the above, and we obtain

$$c = \pm\,\dfrac{2\,(-A + \tfrac{1}{3}A_2{}^2)^{\frac{3}{2}}}{3\,\sqrt{3}} = \dfrac{2}{27}A_2{}^3 - \dfrac{A_2A}{3} + A_0$$

therefore $A_0 = \dfrac{9\,A_2A - 2\,A_2{}^3 \pm 2\,(-3\,A + A_2{}^2)^{\frac{3}{2}}}{27}$

take away from $X_1 = 0$ its second term and we have

$$Y = y^3 + (A - \tfrac{1}{3}A_2{}^2)\,y \pm \dfrac{2\,(-A + \tfrac{1}{3}A_2{}^2)^{\frac{3}{2}}}{3\,\sqrt{3}} = 0\,;\quad \text{this}$$

(prop. I.) contains two equal roots; and therefore the original equation $X_1 = 0$, from which $Y = 0$ was derived, must contain two equal roots.

The roots of $X_1 = 0$ are

$$x = -\frac{A_2}{3} \pm \sqrt{-\frac{A - \frac{1}{3} A_2{}^2}{3}}$$

$$x = -\frac{A_2}{3} \pm \sqrt{-\frac{A - \frac{1}{3} A_2{}^2}{3}}$$

$$x = -\frac{A_2}{3} \mp 2 \sqrt{-\frac{A - \frac{1}{3} A_2{}^2}{3}}$$

or

$$x = -\frac{A_2}{3} \pm \frac{1}{3} (-3 A + A_2{}^2)^{\frac{1}{2}}$$

$$x = -\frac{A_2}{3} \pm \frac{1}{3} (-3 A + A_2{}^2)^{\frac{1}{2}}$$

$$x = -\frac{A_2}{3} \mp \frac{2}{3} (-3 A + A_2{}^2)^{\frac{1}{2}}$$

CHAPTER V.

CUBE ROOTS.

PROPOSITION.

(48.) If the coefficient b in the general equation Y, becomes $= 0$, and c is either positive or negative, fractional or integral, real or imaginary, or consists of either powers or roots, then $Y = y^3 \pm c = 0$, will have the following general values for its roots.

The two equations become—

Proposed equation $\qquad Y = y^3 \pm c = 0$

Equation of differences $Z = z^3 \pm \sqrt{-27\,c^2} = 0$

or $\qquad\qquad\qquad Z = z^3 \pm 3\sqrt{3} \cdot c\,\sqrt{-1} = 0$

The roots of Y for the upper sign will be

$$\frac{c^{\frac{1}{3}}}{2} - \frac{c^{\frac{1}{3}}}{2}\sqrt{-3}, \quad \frac{c^{\frac{1}{3}}}{2} + \frac{c^{\frac{1}{3}}}{2}\sqrt{-3}, \quad -c^{\frac{1}{3}}$$

For the lower sign, $-\dfrac{c^{\frac{1}{3}}}{2} + \dfrac{c^{\frac{1}{3}}}{2}\sqrt{-3}, -\dfrac{c^{\frac{1}{3}}}{2} - \dfrac{c^{\frac{1}{3}}}{2}\sqrt{-3}, +c^{\frac{1}{3}}$

Upper sign of Z, $+c^{\frac{1}{3}}\sqrt{-3}, \dfrac{3\,c^{\frac{1}{3}}}{2} - \dfrac{c^{\frac{1}{3}}}{2}\sqrt{-3}, -\dfrac{3\,c^{\frac{1}{3}}}{2} - \dfrac{c^{\frac{1}{3}}}{2}\sqrt{-3}$

Lower sign of Z, $-c^{\frac{1}{3}}\sqrt{-3}, -\dfrac{3\,c^{\frac{1}{3}}}{2} + \dfrac{c^{\frac{1}{3}}}{2}\sqrt{-3}, \dfrac{3\,c^{\frac{1}{3}}}{2} + \dfrac{c^{\frac{1}{3}}}{2}\sqrt{-3}$

Demonstration.—If any one of the values of Y or Z be substituted for y or z in their respective equations, Y or Z will be reduced to nothing; therefore these values must be roots. Or, the sum of each of the three roots of each equation is $= 0$; the sum of their products taken two and two is $= 0$; and their continued product with changed signs is $=$ the absolute terms of the respective equations; therefore, they are the cube roots of the respective equations Y and Z.

Thus it is proved that every real number has three cube roots, two of which are imaginary: and it is also proved that the differences are all imaginary.

COROLLARY 1.

(49.) If, in equation Y, c is imaginary, and of the form of $c \sqrt{-1}$, then Y will have three imaginary roots, and Z only two which are imaginary, and one real.

The two equations become

$$Y = y^3 \pm c \sqrt{-1} = 0$$

$$Z = z^3 \pm \sqrt{-27 (c \sqrt{-1})^2} = 0$$

or $$y = \mp (c \sqrt{-1})^{\frac{1}{3}}$$

and $$z = \mp \sqrt{3} \cdot c^{\frac{1}{3}}$$

or $$y = \pm c^{\frac{1}{3}} \sqrt{-1}$$

$$z = \mp c^{\frac{1}{3}} \sqrt{3}$$

The roots of Y, by the general formula, are

$$\pm \left(\frac{(c\sqrt{-1})^{\frac{1}{3}}}{2} - \frac{(c\sqrt{-1})^{\frac{1}{3}}}{2} \sqrt{-3} \right), \pm \left(\frac{(c\sqrt{-1})^{\frac{1}{3}}}{2} + \frac{(c\sqrt{-1})^{\frac{1}{3}}}{2} \sqrt{-3} \right), \mp (c\sqrt{-1})^{\frac{1}{3}}$$

The roots of Z are

$$z = \pm \left(c \sqrt{-1}\right)^{\frac{1}{3}} \sqrt{-3} ,$$

$$z = \pm \left\{ \frac{3\left(c\sqrt{-1}\right)^{\frac{1}{3}}}{2} - \frac{\left(c\sqrt{-1}\right)^{\frac{1}{3}}}{2} \sqrt{-3} \right\} ,$$

$$z = \mp \left(\frac{3c\sqrt{-1}^{\frac{1}{3}}}{2} + \frac{\left(c\sqrt{-1}\right)^{\frac{1}{3}}}{2} \sqrt{-3} \right) .$$

But $\pm \left(c \sqrt{-1}\right)^{\frac{1}{3}} \sqrt{-3} = \pm c^{\frac{1}{3}} \sqrt{3}$ which is a real quantity, as deduced directly from the equation Z : all the other roots are imaginary : each of those consisting of two terms, have one term real and the other imaginary ; therefore the corollary is proved.

COROLLARY 2.

(50.) If, in equation Y, c is imaginary and of the form of $a, + c \sqrt{-1}$, then both Y and Z will each have all their roots imaginary.

Demonstration.—The two equations become

$$Y = y^3 \pm \left(a, + c \sqrt{-1}\right) = 0$$

and $\quad Z = z^3 \pm \sqrt{-27\left(a, + c \sqrt{-1}\right)^2} = 0$ which reduce to

$$y = \mp \left(a, + c \sqrt{-1}\right)^{\frac{1}{3}}$$

$$z = \pm \sqrt{3} \cdot \left(a, + c \sqrt{-1}\right)^{\frac{1}{3}} \sqrt{-1}$$

By the general formula the roots of Y and Z become

$$y = \pm \left(\frac{\left(a, + c \sqrt{-1}\right)^{\frac{1}{3}}}{2} - \frac{\left(a, + c \sqrt{-1}\right)^{\frac{1}{3}}}{2} \sqrt{-3} \right) ,$$

$$y = \pm \left(\frac{(a, + c \sqrt{-1})^{\frac{1}{3}}}{2} + \frac{(a, + c \sqrt{-1})^{\frac{1}{3}}}{2} \sqrt{-3} \right),$$

$$y = \mp \ (a, + c \sqrt{-1})^{\frac{1}{3}}$$

$$z = \pm \ (a, + c \sqrt{-1})^{\frac{1}{3}} \sqrt{-3}$$

$$z = \pm \left(\frac{3 (a, + c \sqrt{-1})^{\frac{1}{3}}}{2} - \frac{(a, + c \sqrt{-1})^{\frac{1}{3}}}{2} \sqrt{-3} \right)$$

$$z = \mp \left(\frac{3 (a, + c \sqrt{-1})^{\frac{1}{3}}}{2} + \frac{(a, + c \sqrt{-1})^{\frac{1}{3}}}{2} \sqrt{-3} \right)$$

These roots are all imaginary; therefore the corollary is proved.

(51.) These two corollaries demonstrate that there are two classes of imaginary quantities; one of which has a real cube root; while the other has all of its roots imaginary. It will also be perceived, that if $a,$ and c are any quantities whatever, either real or imaginary, positive or negative, fractional or integral, roots or powers, that the same relations must exist, as have been already demonstrated in the proposition. Or if $c = 0$, both Y and Z become 0; hence the three cube roots and their differences become equal; each being $= 0$.

CHAPTER VI.

PROPOSITION I.

(52.) When the coefficients of the equation of differences of a proposed cubic equation are real, the three roots of the proposed equation are real.

For the demonstration of this proposition see article (40.)

PROPOSITION II.

(53.) *If, in a cubic equation whose second term is wanting, b is real, and if the equation contains three unequal real roots, or three imaginary roots of a simple form, it will admit of a general solution in terms of its coefficients, expressed in the form of a series.*

Demonstration.—First, the equation for real roots will be $Y = y^3 + by + c = 0$; in this b is negative. Assume z equal to one of the differences between two of the roots of Y; let $\dfrac{z}{y} = r$; then $z = ry$, and $y + z = y + ry = (r + 1) y = $ a second root; and $-y - (r + 1) y = -(r + 2) y = $ the remaining root: thus we have the form of the three roots as follows:

$$y, \quad (r + 1) y, \quad -(r + 2) y$$

The sum of the products of these, taken two and two, are

$$\left\{-(r+2)^2 + (r+1)\right\} y^2 = \quad b$$

that is

$$\left\{(r+2)^2 - (r+1)\right\} y^2 = -b$$

therefore

$$y = \pm\sqrt{\frac{-b}{(r+2)(r+1)+1}} \quad \cdots \cdots \quad (1)$$

To find the value of $(r+2)(r+1)$, take the continued product of the three roots, which, when the sign is changed, will equal c:

thus

$$\left\{(r+2)(r+1)\right\} y^3 = c$$

square both members

$$\left\{(r+2)(r+1)\right\}^2 y^6 = c^2$$

also

$$\left\{(r+2)(r+1)+1\right\}^3 y^6 = -b^3$$

divide

$$\frac{\left\{(r+2)(r+1)+1\right\}^3}{\left\{(r+2)(r+1)\right\}^2} = -\frac{b^3}{c^2}$$

by reduction we obtain

$$\left\{(r+2)(r+1)\right\}^2 + \left(\frac{b^3}{c^2}+3\right)(r+2)(r+1) = -3 - \frac{1}{(r+2)(r+1)}$$

therefore

$$(r+2)(r+1) = \frac{1}{2}\left(-\frac{b^3}{c^2}-3\right) \pm \sqrt{\frac{1}{4}\left(-\frac{b^3}{c^2}-3\right)^2 - 3 - \frac{1}{(r+2)(r+1)}}$$

and by substituting the value of $(r+2)(r+1)$ in the denominator of the fraction in the right hand member of equation (1), and adding 1 to both members we have

$$(r+2)(r+1)+1 = 1 + \frac{1}{2}\left(-\frac{b^3}{c^2}-3\right)$$

$$\pm\cfrac{1}{\dfrac{1}{2}\left(-\dfrac{b^3}{c^2}-3\right)-\cfrac{1}{\sqrt{\dfrac{1}{4}\left(-\dfrac{b^3}{c^2}-3\right)^2-3-}\cfrac{1}{\dfrac{1}{2}\left(-\dfrac{b^3}{c^2}-3\right)\pm\sqrt{\dfrac{1}{4}\left(-\dfrac{b^3}{c^2}-3\right)^2}-3-\&c.}}}$$

and by substituting this in equation (1), we have the value of y expressed in a series which is highly convergent, and which, as hereafter to be proved, will be found of great service in obtaining the roots of cubic equations :

$$\text{I.} \quad y = \pm\left\{\cfrac{-b}{1+\dfrac{1}{2}\left(-\dfrac{b^3}{c^2}-3\right)\pm\sqrt{\dfrac{1}{4}\left(-\dfrac{b^3}{c^2}-3\right)^2-3-\cfrac{1}{\dfrac{1}{2}\left(-\dfrac{b^3}{c^2}-3\right)\pm\sqrt{\dfrac{1}{4}\left(-\dfrac{b^3}{c^2}-3\right)^2-3-}\&c.,\&c.\ldots}}}\right\}^{-\frac{1}{2}}$$

It must be remembered, when the roots are real, that b in equation $Y = 0$ is negative, therefore $-b$ and $-\dfrac{b^3}{c^2}$ are positive quantities.

Secondly, the equation for three imaginary roots of a simple form will have b positive and c imaginary : thus $Y = y^3 + by + c \sqrt{-1} = 0$. It will be found, by the same process of demonstration used above, that

$$\left\{ (r+2)\ (r+1) + 1 \right\} y^2 = b$$

and

$$\left\{ (r+2)\ (r+1) \right\} y^3 \sqrt{-1} = + c \sqrt{-1}$$

therefore

$$\frac{\left\{ (r+2)\ (r+1) + 1 \right\}^3}{\left\{ (r+2)\ (r+1) \right\}^2} = \frac{b^3}{c^2}$$

Therefore as both b and $\dfrac{b^3}{c^2}$ are positive and real quantities, they must be substituted in formula I., for $- b$ and $- \dfrac{b^3}{c^2}$ to obtain that part of the root which is to be prefixed to $\sqrt{-1}$: thus the proposition is proved, both for real and for a certain form of imaginary roots.

PROPOSITION III.

(54.) If b is real, and $c = \dfrac{\sqrt{-2\ b^3}}{3 \sqrt{3}}$, and if the equation contains three real or three imaginary roots of a simple form ; then the ratio of $- \dfrac{b^3}{c^2} = - \dfrac{b_{/}^{\,3}}{c_{/}^{\,2}}$; $b_{/}$ and $c_{/}$ being the coefficients of the equation of differences ; and the ratio of each will be equal to 13 · 5.

Demonstration.—The proposed equation will be

$$Y = y^3 + by \pm \frac{\sqrt{-2\ b^3}}{3 \sqrt{3}} = 0$$

and the equation of differences

$$Z = z^3 + 3\,b\,z \pm \sqrt{ - 4\,b^4 - 27 \left(\frac{\sqrt{-2\ b^3}}{3 \sqrt{3}} \right)^2 } = 0$$

that is

$$Z = z^3 + 3bz \pm \sqrt{-2b^3} = 0$$

therefore
$$-\frac{b^3}{\left(\dfrac{\sqrt{-2b^3}}{3\sqrt{3}}\right)^2} = -\frac{(3b)^3}{(\sqrt{-2b^3})^2} = 13\cdot 5 ;$$

that is
$$-\frac{b^3}{c^2} = -\frac{b_{\prime}^3}{c_{\prime}^2} = 13\cdot 5$$

When b is positive, c^2 or $c_{\prime}^2$ is negative; when b is negative, c^2 or $c_{\prime}^2$ is positive.

COROLLARY 1.

(55.) When $-\dfrac{b^3}{c^2} < 13\cdot 5$, $\quad -\dfrac{b_{\prime}^3}{c_{\prime}^2} > 13\cdot 5$; and *vice versa*,

when $-\dfrac{b_{\prime}^3}{c_{\prime}^2} < 13\cdot 5$, $\quad -\dfrac{b^3}{c^2} > 13\cdot 5$

Demonstration.—If c in Y be increased, $c_{\prime}$ in Z will be diminished; but to increase c will diminish the ratio of $-\dfrac{b^3}{c^2}$ below $13\cdot 5$; and as $c_{\prime}$ diminishes, the ratio of $-\dfrac{b_{\prime}^3}{c_{\prime}^2}$ must increase above $13\cdot 5$: and also if $c_{\prime}$ be increased, the ratio of $-\dfrac{b_{\prime}^3}{c_{\prime}^2}$ will fall below $13\cdot 5$; but as $c_{\prime}$ increases c diminishes; therefore the ratio of $-\dfrac{b^3}{c^2}$ rises above $13\cdot 5$.

(56.) By reference to formula I. prop. II., it will be perceived, that the greater the ratio of $-\dfrac{b^3}{c^2}$, the more rapidly the formula converges: and by the corollary in this proposition it is demonstrated that the coefficients of either Y or Z, in all equations which can occur, will furnish a ratio not less than $13\cdot 5$.

COROLLARY 2.

(57.) *When the three roots of the proposed equation* $Y=0$ *are real, the equation will admit of a general solution.*

Demonstration.—The proposed equation will be

$$Y = y^3 + by \pm \frac{\sqrt{-2\,b^3}}{3\sqrt{3}} = 0$$

transform into $Z = z^3 + 3\,b\,z \pm \sqrt{-4\,b^3 - 27\left(\frac{\sqrt{-2\,b^3}}{3\sqrt{3}}\right)^2} = 0$

that is $\qquad Z = z^3 + 3\,b\,z \pm \sqrt{-2\,b^3} = 0$

Multiply the roots of $Y = 0$ by $\sqrt{3}$ and we obtain

$$Y' = y'^3 + 3\,b\,y' \pm \sqrt{-2\,b^3} = 0$$

Thus it will be seen that Z and Y' are identical equations; therefore $z = \sqrt{3}\,y$; but z is one of the differences between the roots of $Y = 0$; therefore the three roots of $Y = 0$ become, article (53),

$$\pm\,y, \quad \pm\,(1 + \sqrt{3})\,y, \quad \mp\,(2 + \sqrt{3})\,y$$

the sum of the products of these taken two and two are

$$(-6 - 3\sqrt{3})\,y^2 = +\,b$$

therefore $\qquad y = \pm \sqrt{-\dfrac{b}{6 + 3\sqrt{3}}}$

therefore we have the three roots as follows :

$$y = \pm \sqrt{-\frac{b}{6 + 3\sqrt{3}}}\,,$$

$$y = \pm\,(1 + \sqrt{3}) \sqrt{-\frac{b}{6 + 3\sqrt{3}}}\,,$$

$$y = \mp\,(2 + \sqrt{3}) \sqrt{-\frac{b}{6 + 3\sqrt{3}}}\,.$$

These roots multiplied by $\sqrt{3}$ will give the three roots of $Z = 0$ or the three differences : thus

$$z = \pm \sqrt{-\frac{b}{2 + \sqrt{3}}},$$

$$z = \pm (1 + \sqrt{3}) \sqrt{-\frac{b}{2 + \sqrt{3}}},$$

$$z = \mp (2 + \sqrt{3}) \sqrt{-\frac{b}{2 + \sqrt{3}}}.$$

When b, in equation $Y = 0$, is negative, the formulas for the roots are positive, and consequently the roots are real; but when b is positive in $Y = 0$, the formulas are negative, and the roots are imaginary.

If the roots in equation $Y = 0$ be multiplied by any quantity whatsoever the same relations will hold, and the equation can be solved in terms of its coefficients: for instance, if $\sqrt{6 + 3\sqrt{3}}$ be used as a multiplier, the three roots of $Y = 0$ will become

$$y = \pm \sqrt{-b},$$

$$y = \pm (1 + \sqrt{3}) \sqrt{-b},$$

$$y = \mp (2 + \sqrt{3}) \sqrt{-b}.$$

and the roots of Z will be

$$z = \pm \sqrt{3} \sqrt{-b},$$

$$z = \pm \sqrt{3} (1 + \sqrt{3}) \sqrt{-b},$$

$$z = \mp \sqrt{3} (2 + \sqrt{3}) \sqrt{-b}.$$

When b in $Y = 0$ is negative, $-b$ in the formulas for the roots is positive.

This same property is also applicable when the three roots are imaginary, and of the form of $a\sqrt{-1}$. It will be shown hereafter that some important consequences grow out of the foregoing properties, enabling us to obtain numerical solutions by a method entirely new.

COROLLARY 3.

$$(58.) \quad \textit{If } A_0 = \frac{9\,A_2\,A - 2\,A_2^3 \pm 2^{\frac{1}{2}}\,(-3\,A + A_2^2)^{\frac{3}{2}}}{27},$$

the general equation becomes

$$X = x^3 + A_2\,x^2 + Ax + \frac{9\,A_2\,A - 2\,A_2^3 \pm 2^{\frac{1}{2}}\,(-3\,A + A_2^2)^{\frac{3}{2}}}{27} = 0,$$

and will admit of a general solution.

Let $X = 0$ be transformed into an equation whose second term shall be absent (31.), and we shall have

$$Y = y^3 + (A - \tfrac{1}{3}A_2^2)\,y \pm \frac{2^{\frac{1}{2}}\,(-A + \tfrac{1}{3}A_2^2)^{\frac{3}{2}}}{3\sqrt{3}} = 0.$$

therefore

$$Z = z^3 + 3\,(A - \tfrac{1}{3}A_2^2)\,z \pm 2^{\frac{1}{2}}\,(-A + \tfrac{1}{3}A_2^2)^{\frac{3}{2}} = 0.$$

For these coefficients substitute b and c, $b_{,}$ and $c_{,}$ and we shall have

$$-\frac{b^3}{c^2} = 13 \cdot 5, \text{ and } -\frac{b_{,}^3}{c_{,}^2} = 13 \cdot 5$$

Hence, by cor. 2, prop. III. the roots of $Y = 0$ will become

$$y = \pm\sqrt{-\frac{A - \tfrac{1}{3}A_2^2}{6 + 3\sqrt{3}}},$$

$$y = \pm\sqrt{-\frac{A - \tfrac{1}{3}A_2^2}{6 + 3\sqrt{3}}}\,(1 + \sqrt{3}),$$

$$y = \mp\sqrt{-\frac{A - \tfrac{1}{3}A_2^2}{6 + 3\sqrt{3}}}\,(2 + \sqrt{3}).$$

To each of these roots add $-\dfrac{A_2}{3}$ and the three roots of $X = 0$ will be obtained: thus

$$x = -\frac{A_2}{3} \pm \sqrt{-\frac{A - \frac{1}{3}A_2^2}{6 + 3\sqrt{3}}},$$

$$x = -\frac{A_2}{3} \pm \sqrt{-\frac{A - \frac{1}{3}A_2^2}{6 + 3\sqrt{3}}}(1 + \sqrt{3}),$$

$$x = -\frac{A_2}{3} \mp \sqrt{-\frac{A - \frac{1}{3}A_2^2}{6 + 3\sqrt{3}}}(2 + \sqrt{3}).$$

PROPOSITION IV.

(59.) If all but two of the fractions whose numerators are unity, in formula I. prop. II. be rejected, the formula will give eight figures of the root, in the most unfavorable equations which can be proposed.

Demonstration.—The most unfavorable equations to be solved by the formula, are those whose coefficients give a ratio $-\dfrac{b^3}{c^2}$ or $-\dfrac{b_{\prime}^3}{c_{\prime}^2}$ as small as $13 \cdot 5$. (See cor. 1, prop. III.) The following is an equation whose coefficients have this ratio :

$$Y = y^3 - 13 \cdot 5\, y \pm 13 \cdot 5 = 0$$

Let a root of Y be calculated by the general formula for such equations, as given in article (57.), last proposition, to ten or twelve places of figures : after which calculate the same, by substituting the coefficients in formula I. proposition II., rejecting all but two of the fractions whose numerators are unity, and it will be found that the first eight figures of the root will be correct.

When either the proposed equation, or the equation of differences, gives a greater ratio for $-\dfrac{b^3}{c^2}$ or $-\dfrac{b_{\prime}^3}{c_{\prime}^2}$ than $13 \cdot 5$, the greater will be the number of figures of the root developed by the formula : for instance, if the proposed equation be $y^3 - 7\,y + 7 = 0$, the ratio $\dfrac{7^3}{7^2} = 7$ is less than $13 \cdot 5$, therefore, the equation of differences, namely, $z^3 - 21\,z + 7 = 0$ must furnish the ratio : hence $\dfrac{\overline{21}^3}{7^2} =$

189; therefore, if 189 be substituted for $-\dfrac{b^3}{c^2}$ in formula I. prop. II.,
and all but one of the fractions whose numerators are unity be rejected,
thirteen figures of the root will be developed ; but if two fractions
be retained, the formula will give upwards of twenty figures of the
root ; and for every additional fraction, the number of figures developed in the root will be increased by eight or ten.   When $-\dfrac{b_i^{\,3}}{c_i^{\,2}}$

is used, instead of $-\dfrac{b^3}{c^2}$, the root obtained will be one of the differences, from which the roots of the proposed equation are easily
derived.

(60.) A few short and simple rules will now be given, for the
reduction and simplification of the general formula, so as in all cases
to obtain the first eight figures of the root.

Let the first eight figures of a root in the equation $Y = 0$, with
as many cyphers annexed as there are remaining figures in the
integral part of the root, be represented by a_8

$$\left[\, 1 + \frac{1}{2}\left(-\frac{b^3}{c^2}-3\right) + \sqrt{\dfrac{1}{4}\left(-\dfrac{b^3}{c^2}-3\right)^2 - 3 - \dfrac{1}{\dfrac{1}{2}\left(-\dfrac{b^3}{c^2}-3\right)+\sqrt{\dfrac{1}{4}\left(-\dfrac{b^3}{c^2}-3\right)^2-3}}}\;\right]^{\frac{1}{2}} \cdot -b$$

If the ratio $-\dfrac{b^3}{c^2}$ is between 30 and 500, then

$$\text{II.} \quad\ldots\ldots\quad a_R = \pm \left[\, 1 + \frac{1}{2}\left(-\frac{b^3}{c^2}-3\right) + \sqrt{\dfrac{1}{4}\left(-\dfrac{b^3}{c^2}-3\right)^2 - 3 - \dfrac{1}{\dfrac{1}{4}\left(-\dfrac{b^3}{c^2}-3\right)+\sqrt{\dfrac{1}{4}\left(-\dfrac{b^3}{c^2}-3\right)^2-3}}}\;\right]^{\frac{1}{2}} \cdot -b$$

If the ratio $-\dfrac{b^3}{c^2}$ is between 500 and 20 000, then

$$\text{III.} \quad\ldots\ldots\quad a_R = \pm \left[\, 1 + \frac{1}{2}\left(-\frac{b^3}{c^2}-3\right) + \sqrt{\dfrac{1}{4}\left(-\dfrac{b^3}{c^2}-3\right)^2 - 3}\;\right]^{-\frac{1}{2}} \cdot -b$$

If the ratio $-\dfrac{b^3}{c^2}$ is between 20 000 and 100 000 000, then

$$\text{IV.} \quad\ldots\ldots\quad a_R = \pm \sqrt{\dfrac{-b}{-\dfrac{b^3}{c^2}-2}}$$

If the ratio $-\dfrac{b^3}{c^2}$ exceeds 100 000 000, then

$$\text{V.} \quad . \quad . \quad . \quad . \quad a_8 = \pm \dfrac{c}{-b}$$

If $-\dfrac{b_{\prime}^{3}}{c_{\prime}^{2}}$ exceeds $13 \cdot 5$, it should be substituted instead of $-\dfrac{b^3}{c^2}$ in the above formulas, and the first eight figures of a root $Z = 0$, (the equation of differences,) will be obtained.

If the root is terminable, its last figure, when found by any of the above formulas, should be increased by unity.

COROLLARY.

(61.) The eight figures of the two remaining roots, can also be easily found, by a process much more simple than by depressing the equation.

Demonstration.—Let d represent the denominator of any one of the above formulas: let $a_8{}'$, $a_8{}''$ represent the first eight figures of each of the remaining roots: then we shall have $a_8{}' = (r + 1)\, a_8$; and $a_8{}'' = - (r + 2)\, a_8$. (See article 53.)

but	$r^2 + 3r + 3 = d$
that is	$r^2 + 3r + 2\tfrac{1}{4} = d - \tfrac{3}{4}$
hence	$r = -\tfrac{3}{2} \pm \sqrt{d - \tfrac{3}{4}}$
therefore	$r + 1 = -\tfrac{1}{2} \pm \sqrt{d - \tfrac{3}{4}}$
and	$-(r + 2) = -\tfrac{1}{2} \mp \sqrt{d - \tfrac{3}{4}}$
therefore	$a'_8 = \left(-\tfrac{1}{2} + \sqrt{d - \tfrac{3}{4}} \right) a_8$
and	$a''_8 = \left(-\tfrac{1}{2} - \sqrt{d - \tfrac{3}{4}} \right) a_8$

PROPOSITION V.

(62.) If the three roots of a cubic equation, $Y = y^3 + by + c = 0$, are real, or if b is real and positive and the three roots imaginary, and the ratio of $-\dfrac{b^3}{c^2}$ is equal to or greater than $13 \cdot 5$, the following formulas I. and II. will be a general solution of the equation, and will furnish any required number of figures of either the roots : and if the ratio of $-\dfrac{b^3}{c^2}$ is less than $13 \cdot 5$, the formulas will solve the equations of differences to any required number of figures.

$$\text{I} \ldots \ldots \; y = a_8 - \frac{c + \left(a_8^2 + b\right) a_8}{3\, a_8^2 + b} - \frac{c + \left(a_{16}^2 + b\right) a_{16}}{3\, a_{16}^2 + b}$$

$$- \frac{c + \left(a_{32}^2 + b\right) a_{32}}{3\, a_{32}^2 + b} - \qquad \cdot \;\cdot$$

$$\text{II} \quad . \; y = a_8 \sqrt{-1} - \frac{c \sqrt{-1} + \left(- a_8^2 + b\right) a_8 \sqrt{-1}}{-3\, a_8^2 + b}$$

$$- \frac{c \sqrt{-1} + \left(- a_{16}^2 + b\right) a_{16} \sqrt{-1}}{-3\, a_{16}^2 + b} -$$

Demonstration.—Let a_8 represent the same value as in the last proposition ; let a_{16}, a_{32}, a_{64}, &c., each represent that part of the root developed by the terms preceding it : thus, the denominator and numerator of the second term, are the values of the last two coefficients of the transformed equation, obtained by diminishing the roots of the proposed equation by a_8 ; therefore by contracted division, this fractional term furnishes eight figures more of the root ; therefore the first two terms of the formula reveal sixteen figures of the root. a_{16} in the third term of the formula represents the sixteen figures of the root, developed by the two preceding terms. Also the denominator and numerator of the third term represent the last two coefficients of the transformed equation, obtained by diminishing the roots of the proposed equation by the sixteen figures before found ; and the third term, by contracted division, furnishes sixteen more figures of the root : thus, the first three terms of the formula reveal thirty-two figures of the root ; and in like manner, the addition of the fourth term will give sixty-four figures of the root ; and so on to any required number.

(63.) The square a_8^2, used in the second term of the formulas in the last art., is already given, being equal, according to the magnitude of the ratios, to some one of the right hand members of the formulas I., II., III., IV. (art. 60.), before the square root is taken : therefore the value of the second term is very readily obtained. It is very rare that over sixteen figures of a root are required ; therefore it is seldom that the value of the third term of either of the formulas will be sought.

To obtain the values of the two remaining roots, substitute a_9' and a_9'' (see art. 61.), in the first and second terms of the formulas, in the last article, and proceed to obtain the other terms as already described.

If the coefficients of the equation of differences are used, instead of those of the proposed equation, the roots obtained, as we have before observed, will be the differences between the roots of the proposed equation, from which the roots of the proposed equation are easily derived.

When the equation contains three imaginary roots of the form of $a_8 \sqrt{-1}$, it will become $Y = y^3 + by + c \sqrt{-1} = 0$; and the general formula will take the form represented in II. of the last art., as will be ascertained by diminishing the roots by $a_8 \sqrt{-1}$, $a_{16} \sqrt{-1}$, &c.

COROLLARY 1.

(64.) If $-\dfrac{b^3}{c^2}$ is less than $13 \cdot 5$, then a_8 or $a_8 \sqrt{-1}$ will represent the first eight figures of one of the differences between the roots of the proposed equation ; and from this developed portion of one of the differences, eight figures or more of each of the remaining five roots of the two equations can be very simply and quickly obtained.

Demonstration.—Let $\pm y'$, $\pm y''$, $\mp y'''$, or $\pm y' \sqrt{-1}$, $\pm y'' \sqrt{-1}$, $\mp y''' \sqrt{-1}$ represent the first eight figures or more of the three roots of the proposed equation Y; let $\pm z'$, $\pm z''$, $\mp z'''$, or $\pm z' \sqrt{-1}$, $\pm z'' \sqrt{-1}$, $\mp z''' \sqrt{-1}$ represent the first eight figures or more of each of the roots of the equation of differences Z; let $z' = a_8$, or $z' \sqrt{-1} = a_8 \sqrt{-1}$, be the eight figures of the root found; let b, = coefficient of Z; then we shall have for eight figures or more of the five remaining roots, the following values :—

For real roots when $\pm z'$ is found.

$$a_{\text{R}} \quad = \quad z' \qquad\qquad .a_{\theta} \quad = \quad - z'$$

$$\frac{c}{\frac{1}{9}\,(3\,a_{\text{R}}{}^2 + b_{,})} = -y''' \qquad\qquad \frac{c}{\frac{1}{9}\,(3\,a_{\theta}{}^2 + b_{,})} = \quad y'''$$

$$\frac{y''' - z'}{2} \quad = \quad y' \qquad\qquad \frac{-y''' + z'}{2} \quad = \quad -y'$$

$$y''' - y' \quad = \quad y'' \qquad\qquad -y''' + y' \quad = \quad -y''$$

$$y''' + y' \quad = \quad z'' \qquad\qquad -y''' - y' \quad = \quad -z''$$

$$-y''' - y' \quad = \quad -z''' \qquad\qquad y''' + y'' \quad = \quad z'''$$

For imaginary roots when $\pm z'\ \sqrt{-1}$ is found.

$$a_{\theta}\ \sqrt{-1} \qquad = \quad z'\ \sqrt{-1}$$

$$\frac{c\ \sqrt{-1}}{\frac{1}{9}\,(-3\,a_{\theta}{}^2 + b_{,})} \quad = -y'''\ \sqrt{-1}$$

$$\frac{y'''\ \sqrt{-1} - z'\ \sqrt{-1}}{2} \quad = \quad y'\ \sqrt{-1}$$

$$y'''\ \sqrt{-1} - y'\ \sqrt{-1} = \quad y''\ \sqrt{-1}$$

$$y'''\ \sqrt{-1} + y'\ \sqrt{-1} = \quad z''\ \sqrt{-1}$$

$$-y'''\ \sqrt{-1} - y''\ \sqrt{-1} = -z'''\ \sqrt{-1}$$

$$a_{\theta}\ \sqrt{-1} \qquad = -z'\ \sqrt{-1}$$

$$\frac{c\ \sqrt{-1}}{\frac{1}{9}\,(-3\,a_{\theta}{}^2 + b_{,})} \quad = \quad y'''\ \sqrt{-1}$$

$$\frac{-y'''\ \sqrt{-1} + z'\ \sqrt{-1}}{2} \quad = -y'\ \sqrt{-1}$$

8

$$- y''' \sqrt{-1} + y' \sqrt{-1} = - y'' \sqrt{-1}$$

$$- y''' \sqrt{-1} - y' \sqrt{-1} = - z'' \sqrt{-1}$$

$$y''' \sqrt{-1} + y'' \sqrt{-1} = \quad z''' \sqrt{-1}$$

In diminishing the roots of the equation of differences by a_8 or $a_8 \sqrt{-1}$, we not only subtract these quantities from themselves, but also from the other two differences ; hence these latter differences become two roots, or rather parts of roots, of the equation of second differences ; and their product is represented by the coefficient $3 a_8{}^2 + b$, or $- 3 a_6{}^2 + b$, of the transformed equation : but the roots of the equation of second differences are thrice the magnitude of the roots of the proposed equation Y; art. (34.) therefore $\frac{1}{9} (3 a_8{}^2 + b,)$ or $\frac{1}{9} (- 3 a_8{}^2 + b,)$ is equal to the product of two roots, or rather parts of roots, of the proposed equation ; therefore c or $c \sqrt{-1}$ divided by this product must be equal to the remaining root y''' or $y''' \sqrt{-1}$, as in the above formula. Therefore z' and y''' being found, the other four roots are by their aid immediately revealed, as in the above simple formulas.

COROLLARY 2.

(65.) If $\pm y'$ be found, the five remaining roots, or rather parts of roots of the two equations, will be expressed by the following simple formulas.

Let $\pm y' = a_8$, or $\pm y' \sqrt{-1} = a_8 \sqrt{-1}$; and let all the other symbols used in this corollary be the same as in the last : and let c, or $c, \sqrt{-1}$ be the absolute term of the equation of differences : then we shall have

For real roots when $\pm y'$ is found.

$$a_8 = y' \qquad\qquad\qquad a_8 = - y'$$

$$\frac{c,}{3 a_8{}^2 + b} = - z''' \qquad\qquad \frac{c,}{3 a_8{}^2 + b} = z'''$$

$$\frac{z''' - 3 y'}{2} = z' \qquad\qquad \frac{- z''' + 3 y'}{2} = - z'$$

$$z''' - z' = z''$$
$$-z''' + z' = -z''$$
$$y' + z' = y''$$
$$-y' - z' = -y''$$
$$-y' - y'' = -y'''$$
$$y' + y'' = y'''$$

For imaginary roots when $\pm\, y'\,\sqrt{-1}$ is found.

$$a_8\,\sqrt{-1} = y'\,\sqrt{-1}$$

$$\frac{c_{,}\,\sqrt{-1}}{-3\,a_8{}^2 + b} = -z'''\,\sqrt{-1}$$

$$\frac{z'''\,\sqrt{-1} - 3\,y'\,\sqrt{-1}}{2} = z'\,\sqrt{-1}$$

$$z'''\,\sqrt{-1} - z'\,\sqrt{-1} = z''\,\sqrt{-1}$$

$$y'\,\sqrt{-1} + z'\,\sqrt{-1} = y''\,\sqrt{-1}$$

$$-y'\,\sqrt{-1} - y''\,\sqrt{-1} = -y'''\,\sqrt{-1}$$

$$a_8\,\sqrt{-1} = -y'\,\sqrt{-1}$$

$$\frac{c_{,}\,\sqrt{-1}}{-3\,a_8{}^2 + b} = z'''\,\sqrt{-1}$$

$$\frac{-z'''\,\sqrt{-1} + 3\,y'\,\sqrt{-1}}{2} = -z'\,\sqrt{-1}$$

$$-z'''\,\sqrt{-1} + z'\,\sqrt{-1} = -z''\,\sqrt{-1}$$

$$-y'\,\sqrt{-1} - z'\,\sqrt{-1} = -y''\,\sqrt{-1}$$

$$y'\,\sqrt{-1} + y''\,\sqrt{-1} = y'''\,\sqrt{-1}$$

In diminishing the roots of the proposed equation by a_8, we subtract this quantity, not only from itself, but also from the other two roots; the result of the last two differences are two roots of the equation of differences, and are expressed in the form of the product $3\,a_8{}^2 + b$, or $-3\,a_8{}^2 + b$ in one of the coefficients of the transformed equation; therefore c, or c, $\sqrt{-1}$, divided by this product, must give the remaining root $\mp z'''$, or $\mp z''' \sqrt{-1}$ of the equation of differences. By the aid of the two roots thus found, the remaining four roots are immediately revealed.

(66.) If it becomes necessary to resort to the equation of differences to find the roots of the proposed equation Y, it is unnecessary to find all the figures of the difference when the number exceeds eight; but carry the process to eight figures which give the value of the first term of the formula I. or II. of this proposition; then, by cor. 1, find eight figures or more of the roots of Y, and proceed to develope the same by the general formula in art. (62.) to any required number of figures. The change from Z to Y requires but little labour, as the value of $a_8{}^2$ in the denominator $\frac{1}{9}(3\, a_8{}^2 + b_{\prime})$, or $\frac{1}{9}(-3\, a_8{}^2 + b_{\prime})$ is already known, as is stated above. See art. (63.)

(67.) In performing the operations of division, multiplication, and extracting the square root, indicated in the formulas for obtaining a_8 see art. (60.); the process should be carried to nine or ten places of figures, so as to ensure exactness in the eighth figure of the root: if, however, the second term of formula I. or II. in art. (62.) is to be used, exactness in the eighth figure is not essential; for if it should happen, through carelessness or in any other way, that the eighth figure is a unit or more too small, the second term of the formula will, in all cases, correct the mistake, and furnish the requisite amount to be added to the eighth figure. Much labour will be saved by using contracted division, contracted multiplication, and such contractions as can be used in extracting the square root; but these short processes will be left to the ingenuity of those who may adopt this method.

(68.) In formulas I. and II., art. (62.), and also in formulas I., II., III., IV., and V. in art. (60.), we have used a_8 as a representation of the number of figures which can be obtained, according to the magnitude of the ratios $-\dfrac{b^3}{c^2}$ or $-\dfrac{b_{\prime}{}^3}{c_{\prime}{}^2}$; but it is not necessary always to obtain just 8, or 16, or 32 figures of a root. For instance, if only 10 figures are wanted, obtain 5 by some one of the formulas in art. (60.), and by the second term of formula I. or II., in art. (62.), 5 more will be added; or if 12 figures of the root are required, either 6 may be obtained by some one of the formulas in art. (60.), and the remaining 6 by the second term of formula I. or II., art. (62.); or 8 figures may be obtained for the first term a_8, and 9 more figures will be added by the second and third terms.

Thus it will be seen that formulas I. and II., in art. (62.), are not only general, but adapted to a multitude of circumstances, according to the number of figures wanted. As a general thing it will be found more expeditious to obtain by art. (60.) only a few figures, say 3, 4, or 5, and obtain the balance by the second, third, &c. terms of formulas I. and II., art. (62.) Or if we choose we can dispense altogether with art. (60.), and obtain a few figures of the root, figure by figure, as hereafter to be explained, and then for the balance of the required figures use the second, third, &c. terms of formulas I. or II., art. (62).

CARDAN'S FORMULA.

PROPOSITION VI.

(69.) If b and c, in the proposed equation, are both real, and the final term of the equation of differences is imaginary, the proposed equation will admit of a general solution.

Demonstration.—Let $Y = y^3 + by + c = 0$ be the proposed equation.

Assume $y = r + r_{,}$. Substitute for y in the proposed equation; thus

$$(r + r_{,})^3 + b (r + r_{,}) + c = 0$$

that is $\qquad r^3 + r_{,}^3 + (3\, r\, r_{,} + b) (r + r_{,}) + c = 0$

Assume $3\, r\, r_{,} + b = 0$, and we shall have

$$r_{,} = -\frac{b}{3\, r}, \text{ and } r_{,}^3 = -\frac{b^3}{27\, r^3}$$

We also have $r^3 + r_{,}^3 + c = 0$

Substitute for the value of $r_{,}^{3}$, and we obtain

$$r^{3} - \frac{b^{3}}{27\, r^{3}} + c = 0$$

that is $r^{6} + c\, r^{3} - \frac{b^{3}}{27} = 0$

Hence $r^{3} = - \frac{c}{2} \pm \sqrt{\left(\frac{c^{2}}{4} + \frac{b^{3}}{27} \right)}$

Therefore $r = \left\{ - \frac{c}{2} \pm \sqrt{\left(\frac{c^{2}}{4} + \frac{b^{3}}{27} \right)} \right\}^{\frac{1}{3}}$

and $r_{,}^{3} = - c - r^{3} = - \frac{c}{2} \mp \sqrt{\left(\frac{c^{2}}{4} + \frac{b^{3}}{27} \right)}$

Therefore $r_{,} = \left\{ - \frac{c}{2} \mp \sqrt{\left(\frac{c^{2}}{4} + \frac{b^{3}}{27} \right)} \right\}^{\frac{1}{3}}$

Therefore $y = r + r_{,} = \left\{ - \frac{c}{2} + \sqrt{\left(\frac{c^{2}}{4} + \frac{b^{3}}{27} \right)} \right\}^{\frac{1}{3}}$

$$+ \left\{ - \frac{c}{2} - \sqrt{\left(\frac{c^{2}}{4} + \frac{b^{3}}{27} \right)} \right\}^{\frac{1}{3}}$$

Thus the value of y is expressed by the sum of two cube roots : but as every quantity has three cube roots, (48.) it is necessary to determine which cube roots are to be used in the present formula.

The cube roots of unity are

$$1,\ \frac{1}{2}\left(- 1 + \sqrt{-3} \right) = a,\ \frac{1}{2}\left(- 1 - \sqrt{-3} \right) = a^{2} \quad \text{See (48.)}$$

If m denotes one of the cube roots of $- \frac{c}{2} + \sqrt{\left(\frac{c^{2}}{4} + \frac{b^{3}}{27} \right)}$, then the other cube roots will be $m\, a$ and $m\, a^{2}$; let n denote one of the cube roots of $- \frac{c}{2} - \sqrt{\left(\frac{c^{2}}{4} + \frac{b^{3}}{27} \right)}$, then the other cube roots are $n\,a$ and $n\,a^2$. The number of pairs of these cube roots will be nine, namely,

$$m + n$$
$$a\ m + a^2\ n$$
$$a^2\ m + a\ n$$
$$m + a^2\ n$$
$$m + a\ n$$
$$a\ m + n$$
$$a\ m + a\ n$$
$$a^2\ m + n$$
$$a^2\ m + a^2\ n$$

But the equation $Y = 0$ can have only three of the above values as roots: the reason of this is because the method of solution requires that $r\,r_{,} = -\dfrac{b}{3}$; and it is this condition which determines the admissible values of the cube roots. For let m and n be such as to satisfy the condition $m\,n = -\dfrac{b}{3}$; in this case we have $r = m$ and $r_{,} = n$ as admissible values: we can also have $r = a\,m$, and $r_{,} = a^2\,n$; for $r\,r_{,} = a\,m \times a^2\,n = a^3\,m\,n = m\,n$; and we can further have $r = a^2\,m$, and $r_{,} = a\,n$; for $r\,r_{,} = a^2\,m \times a\,n = m\,n$: but the product of any other two of the above cube roots will not be equal to $m\,n$; and therefore their sums cannot represent any one of the roots of $Y = 0$; for instance, the product of any pair of the last six values will be either $a\,m\,n = -\dfrac{a\,b}{3}$, or $a^2\,m\,n = -\dfrac{a^2\,b}{3}$, ($a^4\,m\,n$ being equal to $a\,m\,n$.) In the process used in the demonstration, the assumed relation $r\,r_{,} = -\dfrac{b}{3}$ was transformed into $r^3\,r_{,}^3 = -\dfrac{b^3}{27}$; but $-\dfrac{b^3}{27} = -\dfrac{(a\,b)^3}{27} = -\dfrac{(a^2\,b)^3}{27}$; therefore if b, in the proposed equation, is changed into $a\,b$ or $a^2\,b$, the expressions for the value of y would not be changed. Thus the nine values above obtained become the roots of three cubic equations, namely, $y^3 + b\,y + c = 0$, $y^3 + a\,b\,y + c = 0$, and $y^3 + a^2\,b\,y + c = 0$.

The equation of differences of the roots of the proposed equation, is $Z = z^3 + 3\,b\,z \pm \sqrt{-4\,b^3 - 27\,c^2} = 0$. Whenever the final term of this equation is real, the radical quantity $\sqrt{\left(\dfrac{c^2}{4} + \dfrac{b^3}{27}\right)}$ in the expression for y will be imaginary; and whenever the final term of $Z = 0$ is imaginary, the radical quantity $\sqrt{\left(\dfrac{c^2}{4} + \dfrac{b^3}{27}\right)}$ will be real: in the latter case, the square roots of a real quantity can be obtained, and hence the cube roots can be found, and therefore the equation $Y = 0$ will admit of a general solution when the final term of $Z = 0$ is imaginary. The expression for the values of y is called CARDAN'S FORMULA, and is asserted by mathematicians to be a " general algebraical solution" of the cubic equation: but the author does not feel himself warranted in receiving this assertion, as will be seen in the following corollary.

COROLLARY.

(70.) When the final term of the equation $Z = 0$ is real, the

expression $y = \left\{ -\dfrac{c}{2} + \sqrt{\left(\dfrac{c^2}{4} + \dfrac{b^3}{27}\right)} \right\}^{\frac{1}{3}}$

$\qquad + \left\{ -\dfrac{c}{2} - \sqrt{\left(\dfrac{c^2}{4} + \dfrac{b^3}{27}\right)} \right\}^{\frac{1}{3}}$, though algebraically correct, is not a solution.

Demonstration. — The final term of $Z = 0$, namely, $\pm \sqrt{-4\,b^3 - 27\,c^2}$ being real, the expressions for y must be imaginary; and as there is no algebraical method, yet discovered, of obtaining the general value of an imaginary quantity, the two cube roots cannot be found, and, therefore, the expression for y is not, in this case, a *solution*, but merely an *irreducible form*.

Quantities whose values are individually known, or which may be obtained from certain reducible forms, are termed *known* quantities; while quantities which, though individually known, are connected with irreducible forms, are termed *unknown*. Therefore, under these circumstances, the conversion of the original equation into these unknown and irreducible forms, can by no means be admitted as a solution. It is often said, that " *the solution is algebraically correct;* "

but it might, with the same propriety, be said, that " the solution is algebraically correct," when a cubic equation, involving the unknown quantity x, is reduced so that its roots are expressed in terms of the unknown quantity z: the latter is entitled to the term *solution* with as much propriety as the former.

Indeed, CARDAN's formula, instead of being a general solution, is very limited in its applications: it cannot be applied to numerous classes of cubic equations, containing two imaginary roots and one real root; neither will it apply generally to those having three imaginary roots; neither is it applicable to any cubic equation having three real and unequal roots.

EXAMPLE.

(71.) What are the roots of $y^3 + 6y - 20 = 0$.

The second term is positive, therefore the equation contains two imaginary and one real root. Also $b = 6$ and $c = -20$; substitute these numbers in the formula, thus

$$ y = (10 + \sqrt{108})^{\frac{1}{3}} + (10 - \sqrt{108})^{\frac{1}{3}} $$

and $(10 + \sqrt{108})^{\frac{1}{3}} = 2 \cdot 782 \ldots$, and $(10 - \sqrt{108})^{\frac{1}{3}} = - \cdot 732 \ldots$, the decimal part of each of these cube roots may be assumed to be equal, though of opposite signs; therefore $y = 2$ may be assumed to be one of the roots, which by trial will be found to be the case.

Depress the equation by dividing by the factor $y - 2 = 0$, and we shall have $y^3 + 6y - 20 = (y - 2)(y^2 + 2y + 10)$; the quadratic equation

$$ y^2 + 2y + 10 = 0 $$

furnishes the other two roots, namely, $-1 \pm 3\sqrt{-1}$

THE AUTHOR'S FORMULA.

PROPOSITION VII.

(72.) *If, in the proposed equation, b is real and c imaginary, and the coefficients of the equation of differences are real, the proposed equation will admit of a general solution.*

Demonstration.—Let $Y = y^3 + b\,y + c\,\sqrt{-1} = 0$ be the proposed equation: transform this into an equation, the differences of whose roots will be the roots of $Y = 0$: thus

$$Z_{,} = z_{,}^3 + \frac{b}{3}z_{,} + \frac{\sqrt{-4\,b^3 + 27\,c^2}}{27} = 0 \quad \text{See art. (35.)}$$

Assume the roots of $Y = 0$ to be $+ 2\,v\,\sqrt{-1}$, $+ 3\,x - v\,\sqrt{-1}$, $- 3\,x - v\,\sqrt{-1}$, then the roots of $Z_{,}$ will be $+ x - v\,\sqrt{-1}$, $+ x + v\,\sqrt{-1}$, $- 2\,x$.

As $+ 2\,v\,\sqrt{-1}$ is the difference between two of the roots of $Z_{,} = 0$, it may be represented thus $y = 2\,v\,\sqrt{-1} = (r_{,} - r)\,\sqrt{-1}$; substitute for y in equation $Y = 0$, and we have

$$- (r_{,} - r)^3\,\sqrt{-1} + b\,(r_{,} - r)\,\sqrt{-1} + c\,\sqrt{-1} = 0$$

that is
$$(r_{,} - r)^3 - b\,(r_{,} - r) - c = 0$$

Hence
$$r_{,}^3 - r^3 + (- 3\,r_{,}\,r - b)\,(r_{,} - r) - c = 0$$

Assume
$$- 3\,r_{,}\,r - b = 0$$

then
$$r = - \frac{b}{3\,r_{,}}\,; \text{ and } - r^3 = + \frac{b^3}{27\,r_{,}^3}$$

thus we have $r_{,}^3 - r^3 - c = 0$; and $- r^3 = c - r_{,}^3$

substitute for $- r^3$, and we have

$$r_{,}^3 + \frac{b^3}{27\,r_{,}^3} - c = 0$$

that is
$$r_{,}^6 - c\,r_{,}^3 + \frac{b^3}{27} = 0$$

Hence
$$r_i{}^3 = \frac{c}{2} \pm \sqrt{\left(\frac{c^2}{4} - \frac{b^3}{27}\right)}$$

and
$$-r^3 = c - r_i{}^3 = \frac{c}{2} \mp \sqrt{\left(\frac{c^2}{4} - \frac{b^3}{27}\right)}$$

Therefore
$$r_i \sqrt{-1} = \left\{\frac{c}{2} \pm \sqrt{\left(\frac{c^2}{4} - \frac{b^3}{27}\right)}\right\}^{\frac{1}{3}} \sqrt{-1}$$

and
$$-r \sqrt{-1} = \left\{\frac{c}{2} \mp \sqrt{\left(\frac{c^2}{4} - \frac{b^3}{27}\right)}\right\}^{\frac{1}{3}} \sqrt{-1}$$

Therefore $y = (r_i - r) \sqrt{-1} = \left\{\frac{c}{2} + \sqrt{\left(\frac{c^2}{4} - \frac{b^3}{27}\right)}\right\}^{\frac{1}{3}} \sqrt{-1}$

$$+ \left\{\frac{c}{2} - \sqrt{\left(\frac{c^2}{4} - \frac{b^3}{27}\right)}\right\}^{\frac{1}{3}} \sqrt{-1}$$

(73.) This formula differs in two of its features from CARDAN's formula. First, when b is positive in the equation $Y = 0$, $-\frac{b^3}{27}$ in my formula is negative, but the same term in CARDAN's formula is positive: and when c in the equation $Y = 0$ is positive, the term $\frac{c}{2}$ remains, in my formula, positive, but in CARDAN's formula the term is negative: or whatever may be the signs prefixed to b and c in the proposed equation $Y = 0$, the signs prefixed to these terms are opposite in the two formulas. Second, the two cube roots in the author's formula are imaginary, when the two cube roots in CARDAN's formula are real.

(74.) When $c \sqrt{-1}$ is positive, the root will be positive; and when $c \sqrt{-1}$ is negative, the root will be negative. Also when b is negative $-\frac{b^3}{27}$ is a positive quantity; in this case, both terms of the radical will be positive: but when b is positive, and $\frac{c^2}{4} < \frac{b^3}{27}$, the radical $\sqrt{\left(\frac{c^2}{4} - \frac{b^3}{27}\right)}$ becomes imaginary; and the two cube roots become irreducible; and therefore the equation becomes also irreducible by this formula; all equations of this latter class will be found to contain three imaginary roots.

Let the root $+ 2 v \sqrt{-1}$ which has been found, be represented by $+ 2 a \sqrt{-1}$, then the assumed roots become $+ 2 a \sqrt{-1}$, $3 x - a \sqrt{-1}$, $- 3 x - a \sqrt{-1}$; the values of the two remaining roots can be determined by a much more expeditious and simple process than by depressing the equation; thus

$$(3 x - a \sqrt{-1})(- 3 x - a \sqrt{-1}) = \frac{- c \sqrt{-1}}{2 a \sqrt{-1}} = - \frac{c}{2 a}$$

that is

$$9 x^2 = \frac{c}{2 a} - a^2$$

$$3 x = \pm \sqrt{\left(\frac{c}{2 a} - a^2\right)}$$

therefore the values of the three roots of Y will be as follows:

$$y = \quad 2 a \sqrt{-1},$$

$$y = \quad \sqrt{\left(\frac{c}{2 a} - a^2\right)} - a \sqrt{-1},$$

$$y = - \sqrt{\left(\frac{c}{2 a} - a^2\right)} - a \sqrt{-1}.$$

The roots of Z are

$$z = \quad \frac{1}{3} \sqrt{\left(\frac{c}{2 a} - a^2\right)} - a \sqrt{-1},$$

$$z = \quad \frac{1}{3} \sqrt{\left(\frac{c}{2 a} - a^2\right)} + a \sqrt{-1},$$

$$z = - \frac{2}{3} \sqrt{\left(\frac{c}{2 a} - a^2\right)}.$$

If all the signs of these six roots, with the exception of those under the radical sign, be changed, they will become the roots of Y and Z, when their final terms are negative.

From the relations between Y and Z, , it may be observed that thrice the roots of Z, are equal to the differences between the roots of Y; and also, as stated above, (72.) the roots of Y are the differences of the roots of $Z_{,}$.

The roots of $Z_{,} = 0$ are generally obtained by CARDAN's formula; but they can also be found, by first finding from my formula the roots of $Y = 0$, and then taking one-third of three of their differences, as directed in art. (82.)

From an examination of these roots, another curious property is observed, namely,

When $\frac{c}{2\,a} = a^2$, one of the roots of $Z_{,}$ becomes $= 0$; and the other two roots are $-a\sqrt{-1}$ and $+a\sqrt{-1}$; while two of the roots of Y become equal. When $\frac{c}{2\,a} < a^2$, the value of x becomes imaginary; therefore all the roots of both Y and $Z_{,}$ will be imaginary: but these may be considered of a simple form, because all the terms of each root are imaginary; that is, no terms of real quantities enter into their composition. When these expressions, or the values of y become imaginary, the radical quantity $\sqrt{\left(\frac{c^2}{4} - \frac{b^3}{27}\right)}$ also becomes imaginary; so that the value of $2\,a$ cannot be determined by the formula.

EXAMPLE.

(75.) Required the roots of the equation

$$y^3 + 18\,y + 108\,\sqrt{-1} = 0$$

Here $b = 18$; and $c = +108$; thus

$$y = (54 + \sqrt{2700})^{\frac{1}{3}}\sqrt{-1} + (54 - \sqrt{2700})^{\frac{1}{3}}\sqrt{-1}$$

By numerical operation

$$(54 + \sqrt{2700})^{\frac{1}{3}} = 4 \cdot 732 \ldots\ldots, \text{ and}$$

$$(54 - \sqrt{2700})^{\frac{1}{3}} = 1 \cdot 268 \text{ nearly.}$$

Therefore $y = 4 \cdot 732 \ldots \sqrt{-1} + 1 \cdot 268\,\sqrt{-1} = 6\,\sqrt{-1} = 2 \times 3\,\sqrt{-1} = 2\,a\,\sqrt{-1}$

The other two roots may be obtained by the formula for the roots of Y in art. (74.); thus

$$y = + \sqrt{\left(\frac{108}{2 \times 3} - 3^2\right)} - 3\sqrt{-1} = \quad 3 - 3\sqrt{-1}$$

$$y = - \sqrt{\left(\frac{108}{2 \times 3} - 3^2\right)} - 3\sqrt{-1} = -3 - 3\sqrt{-1}$$

If the final term, in the above example, had been minus instead of plus, then we should have had

$$y = (-54 + \sqrt{2700})^{\frac{1}{3}}\sqrt{-1} + (-54 - \sqrt{2700})^{\frac{1}{3}}\sqrt{-1}$$

Hence $(-54 + \sqrt{2700})^{\frac{1}{3}} = -1 \cdot 268$ nearly, and

$$(-54 - \sqrt{2700})^{\frac{1}{3}} = -4 \cdot 732 \ . \ . \ . \ . \ .$$

Therefore $y = -1 \cdot 268 \sqrt{-1} - 4 \cdot 732 \ . \ . \ . \ . \ \sqrt{-1} =$

$$-6\sqrt{-1} = -2 \times 3 \sqrt{-1} = -2 a \sqrt{-1}$$

$$y = - \sqrt{\left(\frac{-108}{-2 \times 3} - 3^2\right)} + 3\sqrt{-1} = -3 + 3\sqrt{-1}$$

$$y = \sqrt{\left(\frac{-108}{-2 \times 3} - 3^2\right)} + 3\sqrt{-1} = \quad 3 + 3\sqrt{-1}$$

One-third of the three differences (32.) of these roots will be

$-1 \pm 3\sqrt{-1}$ and 2 which are the roots of $y^3 + 6y - 20 = 0$ See (71.)

By this last example, it will more fully be perceived, how this new method accomplishes the same results as CARDAN's formula; and at the same time reveals the three imaginary roots of an equation.

CHAPTER VII.

(76.) As all complete biquadratic equations can be easily transformed into others whose second term shall be absent, art. (31.) our investigations will be more particularly directed to this latter class. We shall first give a " general solution " of the equation, agreeing in some respects with *Descartes' Solution* ; but in other respects quite different, resulting in an auxiliary cubic equation having no second term.

"GENERAL SOLUTION" OF THE BIQUADRATIC EQUATION.

PROPOSITION I.   PROBLEM.

(77.) Required, the roots of the biquadratic equation, expressed in terms of its coefficients.

Let $X_{,} = x^4 + q\,x^2 + r\,x + s = 0$ be the general equation. Assume

$$X_{,} = x^4 + q\,x^2 + r\,x + s =$$

$$\left(x^2 + \sqrt{\frac{y - 2\,q}{3}}\,.\,x + f\right)\left(x^2 - \sqrt{\frac{y - 2\,q}{3}}\,.\,x + y\right)$$

Now if the quantities y, f, and g can be found, we shall have the values of the four roots of $X_{,}$, expressed in two quadratic factors.   Multiply together these factors, and equate the coefficients of like powers of x in both members of the equation ; thus

$$g + f - \frac{y - 2\,q}{3} = q \,,\, (g - f)\sqrt{\frac{y - 2\,q}{3}} = r \,,\, gf = s$$

that is

$$g + f = q + \frac{y - 2\,q}{3} \,,\quad g - f = \frac{r}{\sqrt{\dfrac{y - 2\,q}{3}}} \,,\, gf = s$$

From the first two equations g and f are found in terms of y; substitute these values in the third equation, and we have

$$\left(q + \frac{y - 2\,q}{3} + \frac{r}{\sqrt{\dfrac{y - 2\,q}{3}}} \right)\left(q + \frac{y - 2\,q}{3} - \frac{r}{\sqrt{\dfrac{y - 2\,q}{3}}} \right) = 4\,s$$

By reducing this equation we obtain

$$\text{I} \ . \ . \ \left(\frac{y - 2\,q}{3} \right)^3 + 2\,q \left(\frac{y - 2\,q}{3} \right)^2 + (q^2 - 4\,s) \left(\frac{y - 2\,q}{3} \right) - r^2 = 0$$

Multiply the roots of I. by three, and take away the second term; and by reducing still further we have

$$\text{II} \ . \ . \ . \ y^3 - 3\,(q^2 + 12\,s)\,y - 2\,q^3 + 72\,q\,s - 27\,r^2 = 0$$

Thus we have arrived at a cubic equation with the second term absent: from this, y can be found (69.) by CARDAN's formula; after which g and f become known; and therefore all the coefficients of the two quadratic factors in equation X, become known; and as the product of these quadratic factors is equal to nothing, each factor can be equated with nothing; and the four roots thus obtained will be the four roots of the biquadratic equation X, , expressed in terms of its coefficients : thus the problem is solved.

(78.) Both equations I. and II., obtained by our investigations of this proposition, will, in subsequent researches, be found very useful, especially equation II., which, lacking the second term, is peculiarly adapted to many investigations to which the other form is not so well suited : therefore we shall denominate equation II. as the *general auxiliary cubic equation* ; and its equation of differences will be called *auxiliary cubic equation of differences*.

PROPOSITION II.

(79.) The four roots of the biquadratic equation can be expressed in the terms of the roots of cubic equation I., prop. I.

Demonstration.—Let y', y'', and y''' be the roots of the auxiliary cubic equation, then we shall have

$$y' + y'' + y''' = 0$$

therefore $\quad \dfrac{y'-2q}{3} + \dfrac{y''-2q}{3} + \dfrac{y'''-2q}{3} = -2q$; this,

when the sign is changed, is equal to the coefficient of the second term of equation I., Prop. I. ; we also have

$$\frac{y'-2q}{3} \times \frac{y''-2q}{3} \times \frac{y'''-2q}{3} = r^2$$

that is $\quad \sqrt{\dfrac{y'-2q}{3}} \times \sqrt{\dfrac{y''-2q}{3}} \times \sqrt{\dfrac{y'''-2q}{3}} = r$

and $\quad \sqrt{\dfrac{y''-2q}{3}} \times \sqrt{\dfrac{y'''-2q}{3}} = -\dfrac{r}{\sqrt{\dfrac{y'-2q}{3}}}$

therefore $x^2 + \sqrt{\dfrac{y-2q}{3}} \cdot x + f$, which is one of the quadratic

factors, used in the last proposition, becomes equal to

$$x^2 + \sqrt{\frac{y'-2q}{3}} \cdot x + \frac{1}{2} \left(q + \frac{y'-2q}{3} - \frac{r}{\sqrt{\frac{y'-2q}{3}}} \right)$$

$$= x^2 + \sqrt{\frac{y'-2q}{3}} \cdot x + \frac{1}{2} \left(\frac{y'-2q}{3} - \frac{y'-2q}{2\times3} - \frac{y''-2q}{2\times3} \right.$$

$$\left. - \frac{y'''-2q}{2\times3} - \sqrt{\frac{y''-2q}{3}} \times \sqrt{\frac{y'''-2q}{3}} \right)$$

$$= x^2 + \sqrt{\frac{y'-2q}{3}} \cdot x + \frac{1}{4} \left(\frac{y'-2q}{3} - \frac{y''-2q}{3} - \frac{y'''-2q}{3} \right.$$

$$\left. - 2\sqrt{\frac{y''-2q}{3}} \times \sqrt{\frac{y'''-2q}{3}} \right) = 0$$

therefore, extracting the square root we find the two values of x which represent by $x_{,}$ and $x_{,,}$.

$$x_{,} = \frac{1}{2} \left(-\sqrt{\frac{y'-2q}{3}} - \sqrt{\frac{y''-2q}{3}} - \sqrt{\frac{y'''-2q}{3}} \right),$$

$$x_{,,} = \frac{1}{2} \left(-\sqrt{\frac{y'-2q}{3}} + \sqrt{\frac{y''-2q}{3}} + \sqrt{\frac{y'''-2q}{3}} \right).$$

In a similar manner we obtain from $x^2 - \sqrt{\dfrac{y - 2\,q}{3}} \cdot x + y = 0$

$$x_{,,,} = \frac{1}{2}\left(\sqrt{\frac{y' - 2\,q}{3}} - \sqrt{\frac{y'' - 2\,q}{3}} + \sqrt{\frac{y''' - 2\,q}{3}} \right),$$

$$x_{,,,,} = \frac{1}{2}\left(\sqrt{\frac{y' - 2\,q}{3}} + \sqrt{\frac{y'' - 2\,q}{3}} - \sqrt{\frac{y''' - 2\,q}{3}} \right).$$

Thus the proposition is demonstrated.

(80.) By an examination of these four biquadratic roots, we find them connected with the roots of the cubic equation i. prop. i. in a very remarkable manner : that is, the square of the sum of any two biquadratic roots is equal to a root of equation i.; thus

$$\left.\begin{aligned}
(x_, + x_{,,})^2 &= (x_{,,,} + x_{,,,,})^2 = \frac{y' - 2\,q}{3}\\[2ex]
(x_, + x_{,,,})^2 &= (x_{,,} + x_{,,,,})^2 = \frac{y'' - 2\,q}{3}\\[2ex]
(x_, + x_{,,,,})^2 &= (x_{,,} + x_{,,,})^2 = \frac{y''' - 2\,q}{3}
\end{aligned}\right\} \quad \dots\dots A$$

Also the sum of any two biquadratic roots is equal to the square root of one of the roots of equation i., prop. i.; and also equal to the coefficient of x in the two quadratic factors of equation $X_,$.

that is　　$x_, + x_{,,}\ ,\ x_{,,,} + x_{,,,,} = \mp\sqrt{\dfrac{y' - 2\,q}{3}}$

$$x_, + x_{,,,}\ ,\ x_{,,} + x_{,,,,} = \mp\sqrt{\frac{y'' - 2\,q}{3}}$$

$$x_, + x_{,,,,}\ ,\ x_{,,} + x_{,,,} = \mp\sqrt{\frac{y''' - 2\,q}{3}}$$

The coefficient of the second term of $X_,$ being nothing, the sum of the four biquadratic roots is equal to nothing.

that is

$$\pm (x_, + x_{,,}) \pm (x_{,,,} + x_{,,,,}) = \mp \sqrt{\frac{y' - 2q}{3}} \pm \sqrt{\frac{y' - 2q}{3}} = 0$$

$$\pm (x_, + x_{,,,}) \pm (x_{,,} + x_{,,,,}) = \mp \sqrt{\frac{y'' - 2q}{3}} \pm \sqrt{\frac{y'' - 2q}{3}} = 0$$

$$\pm (x_, + x_{,,,,}) \pm (x_{,,} + x_{,,,}) = \mp \sqrt{\frac{y''' - 2q}{3}} \pm \sqrt{\frac{y''' - 2q}{3}} = 0$$

Thus it is perceived, that the sum of any two roots is equal in magnitude and opposite in sign to the remaining two roots.

(81.) When the coefficient of x in $x^2 + \sqrt{\frac{y - 2q}{3}} \cdot x + f = 0$, or $x^2 - \sqrt{\frac{y - 2q}{3}} \cdot x + g = 0$, is found, the positive value of $\pm \sqrt{\frac{y - 2q}{3}}$ may be substituted in the second equation, and the negative value in the first, and it will make no difference in the results of the solution ; the only effect produced is merely the interchange of the values of f and g.

PROPOSITION III.

(82.) 1. — If the roots of the biquadratic equation are all real, the roots of cubic equation I., prop. I., will all be real and positive.

2. — If the coefficients of the biquadratic equation are all real, and its roots all imaginary, the roots of the cubic equation I., prop. I., will all be real, but two will be negative and one positive.

3. — If the coefficients of the biquadratic equation are all real, and two of its roots real and two imaginary, the cubic equation I. will have one real and two imaginary roots, except in case of equal roots, when its three roots will be real, and two of them negative.

Demonstration 1. — By reference to A in proposition II., it will be seen, that the square of the sum of any two biquadratic roots is equal to a root of cubic equation I., prop. I. Now if these biquadratic roots are real, the squares of any two pairs of such roots must, not only be real, but also positive.

2. — If the coefficients of the biquadratic equation are all real, and its four roots imaginary, they must necessarily be of the forms.

$$x_{,} = + a + a_{,} \sqrt{-1} \qquad x_{,,} = a - a_{,} \sqrt{-1}$$

$$x_{,,,} = - a + a_{2} \sqrt{-1} \qquad x_{,,,,} = - a - a_{2} \sqrt{-1}$$

hence $\quad x_{,} + x_{,,} = 2\,a$ and $x_{,,,} + x_{,,,,} = -2\,a$

$$x_{,} + x_{,,,} = (a_{,} + a_{2}) \sqrt{-1} \text{ and } x_{,,} + x_{,,,,} = -(a_{,} + a_{2}) \sqrt{-1}$$

$$x_{,} + x_{,,,,} = (a_{,} - a_{2}) \sqrt{-1} \text{ and } x_{,,} + x_{,,,} = -(a_{,} - a_{2}) \sqrt{-1}$$

therefore $(x_{,} + x_{,,})^{2} = (x_{,,,} + x_{,,,,})^{2} = + 4\,a^{2}$

$$(x_{,} + x_{,,,})^{2} = (x_{,,} + x_{,,,,})^{2} = -(a_{,} + a_{2})^{2}$$

$$(x_{,} + x_{,,,,})^{2} = (x_{,,} + x_{,,,})^{2} = -(a_{,} - a_{2})^{2}$$

By reference to A in prop. II. it will be seen, that the quantities represented by the squares of these sums, are the roots of equation I. prop. I. ; but these quantities are real and two of them are negative ; and therefore the second part of the proposition is proved.

3. — If the coefficients of the biquadratic equation are real, and two of its roots are real and two imaginary, then the four roots must necessarily have the following forms :—

$$x_{,} = a + a_{2} \sqrt{-1}, \qquad x_{,,} = a - a_{2} \sqrt{-1}$$

$$x_{,,,} = - a + a_{,} \qquad , \qquad x_{,,,,} = - a - a_{,}$$

hence

$$x_{,} + x_{,,} = 2\,a \qquad \text{and } x_{,,,} + x_{,,,,} = -2\,a$$

$$x_{,} + x_{,,,} = a_{,} + a_{2} \sqrt{-1} \text{ and } x_{,,} + x_{,,,,} = -(a_{,} + a_{2} \sqrt{-1})$$

$$x_{,,} + x_{,,,} = a_{,} - a_{2} \sqrt{-1} \text{ and } x_{,} + x_{,,,,} = -(a_{,} - a_{2} \sqrt{-1})$$

therefore

$$(x_{,} + x_{,,})^{2} = (x_{,,,} + x_{,,,,})^{2} = 4\,a^{2}$$

$$(x_{,} + x_{,,,})^{2} = (x_{,,} + x_{,,,,})^{2} = (a_{,} + a_{2} \sqrt{-1})^{2}$$

$$(x_{,,} + x_{,,,})^{2} = (x_{,} + x_{,,,,})^{2} = (a_{,} - a_{2} \sqrt{-1})^{2}$$

These are the roots of the cubic equation I. prop. I., as will be seen by reference to A, prop. II.; and when $a_{,}$ is not zero, two of these roots must be imaginary; but when $a_{,}$ is zero, the three roots will be real, and two of them equal and negative; therefore all that was asserted in the proposition is demonstrated.

(83.) For the convenience of reference, let the auxiliary cubic equation, the auxiliary equation of differences, and the cubic equation I., prop I., be represented respectively by Y, Z, and $Y_{,}$; thus

$$Y = y^3 - 3\left(q^2 + 12\,s\right)y + 72\,q\,s - 2\,q^3 - 27\,r^2 = 0$$

$$Z = z^3 - 9\left(q^2 + 12\,s\right) z$$

$$\pm \sqrt{-4\left(-3\left(q^2 + 12s\right)\right)^3 - 27\left(72\,q\,s - 2\,q^3 - 27\,r^2\right)^2} = 0$$

$$Y_{,} = \left(\frac{y - 2\,q}{3}\right)^3 + 2\,q\left(\frac{y - 2\,q}{3}\right)^2 + \left(q^2 - 4\,s\right)\left(\frac{y - 2\,q}{3}\right) - r^2 = 0$$

Also let the coefficients of Y be represented by b and c; and those of Z by $b_{,}$ and $c_{,}$; thus

$$Y = y^3 + by + c = 0$$
$$Z = z^3 + b_{,} z + c_{,} = 0$$

(84.) Having determined the properties of these three equations in their relations to the biquadratic equation, a proposition of considerable importance suggests itself, namely, to determine by some simple process, the nature of the roots of a biquadratic equation when its coefficients are real, under all circumstances which can occur. This is a proposition which has engaged the attention of mathematicians during the present and past centuries, and in general has resulted in numerous cumbersome theories, calculated by their enormous amount of labour, to greatly discourage the young student in his mathematical progress.

PROPOSITION IV.

(85.) 1. — If all the coefficients of a proposed biquadratic equation are real, and the final term $c_{,}$ of the auxiliary cubic equation of differences Z, is either real or equal to nothing, or if b and c are each equal to nothing, and the coefficients of $Y_{,}$ are alternately positive and negative, then all the roots of the biquadratic equation are real.

2. — If c, is real, and the coefficients of Y, are not alternately positive and negative, the four roots of the proposed biquadratic equation are imaginary.

3. — If c, is zero, and the coefficients of Y, are not alternately positive and negative, two roots of the proposed biquadratic equation are real and equal, and two roots are imaginary.

4. — If c, is imaginary, two roots of the proposed biquadratic equation are real, and two roots imaginary,

Demonstration 1. — If the coefficients of Y, are alternately positive and negative, then, by art. (21.), its three roots must be real and positive, or else two of its roots are imaginary: If two of its roots are imaginary, the equation of differences will have three imaginary roots, and hence, c, will be imaginary; but c, by hypothesis is real; therefore the three roots of Y, are real and positive, therefore, by prop. III., the four roots of the biquadratic equation are real.

2. — If c, is real, Y, cannot have imaginary roots; and when its coefficients are not alternately positive and negative, it must, by prop. III., have one positive and two negative roots; and therefore by the same prop. all the roots of the biquadratic equation are imaginary.

3. — If c, is zero, one of the differences between the roots of Y , and consequently of Y, will be nothing; therefore Y, will have two equal roots, and if its coefficients are not alternately positive and negative, these two equal roots will by prop. III. be negative; and therefore by the same prop. the biquadratic equation must have two equal real roots and two imaginary roots.

4. — If c, is imaginary, two of the roots of Y, will be imaginary; and therefore, by prop. III., two of the roots of the proposed biquadratic equation will be real, and two imaginary.

These are all the cases which can occur, when the coefficients of the biquadratic equation are real.

PROPOSITION V.

(86.) *When* $r = \pm \sqrt{\left(\dfrac{- 2\, q^3 + 72\, q\, s \mp 2\, (q^2 + 12\, s)^{\frac{3}{2}}}{27} \right)}$ *in the equation* $X, = x^4 + q\, x^2 + r\, x + s = 0$, *the equation will contain two or more equal roots, and will admit of a general solution.*

Demonstration.— Let the equation

$$X, = x^4 + q\,x^2 \pm \sqrt{\left(\frac{-2\,q^3 + 72\,q\,s \mp 2\,(q^2 + 12\,s)^{\frac{3}{2}}}{27}\right)}\,x + s = 0$$

be transformed into a cubic equation, according to art. (77.) ; and let $y = \frac{y}{3}$; then equation I., art. (77.), will become

$$Y, = \left(y - \frac{2\,q}{3}\right)^3 + 2\,q\left(y - \frac{2\,q}{3}\right)^2 + (q^2 - 4\,s)\left(y - \frac{2\,q}{3}\right)$$

$$+ \frac{2\,q^3 - 72\,q\,s \pm 2\,(q^2 + 12\,s)^{\frac{3}{2}}}{27} = 0$$

Transform this equation into another whose second term shall be absent, and we obtain

$$Y = y^3 - (\tfrac{1}{3}\,q^2 + 4\,s)\,y \pm \frac{2\,(q^2 + 12\,s)^{\frac{3}{2}}}{27} = 0$$

Transform $Y = 0$ into an equation of differences : thus

$$Z = z^3 - 3\,(\tfrac{1}{3}\,q^2 + 4\,s)\,z$$

$$\pm \sqrt{-4\,[-(\tfrac{1}{3}\,q^2 + 4\,s)^3] - 27\left(\frac{2\,(q^2 + 12\,s)^{\frac{3}{2}}}{27}\right)^2} = 0$$

that is $Z = z^3 - 3\,(\tfrac{1}{3}\,q^2 + 4\,s)\,z = 0$

Hence one of the differences $z = 0$; therefore $Y = 0$ has two equal roots. The roots of Y, by art. (42), áre

$$\pm \sqrt{\frac{\tfrac{1}{3}\,q^2 + 4\,s}{3}},\ \pm \sqrt{\frac{\tfrac{1}{3}\,q^2 + 4\,s}{3}},\ \mp 2\sqrt{\frac{\tfrac{1}{3}\,q^2 + 4\,s}{3}}\,.$$

therefore, the roots of equation Y, in this proposition are (47.)

$$\left(y - \frac{2\,q}{3}\right) = -\frac{2\,q}{3} \pm \sqrt{\frac{\tfrac{1}{3}\,q^2 + 4\,s}{3}},$$

$$\left(y - \frac{2\,q}{3}\right) = -\frac{2\,q}{3} \pm \sqrt{\frac{\tfrac{1}{3}\,q^2 + 4\,s}{3}},$$

$$\left(y - \frac{2\,q}{3}\right) = -\frac{2\,q}{3} \mp 2\sqrt{\frac{\tfrac{1}{3}\,q^2 + 4\,s}{3}}.$$

And the four roots of $X_{,}$, by art. (78.) become

$$x_{,} = \frac{1}{2}\left\{ -2\sqrt{\left(-\frac{2\,q}{3} \pm \sqrt{\frac{\frac{1}{3}\,q^2 + 4\,s}{3}} \right)} \right.$$

$$\left. -\sqrt{\left(-\frac{2\,q}{3} \mp 2\sqrt{\frac{\frac{1}{3}\,q^2 + 4\,s}{3}} \right)} \right\},$$

$$x_{,,} = \frac{1}{2}\left\{ \sqrt{\left(-\frac{2\,q}{3} \mp 2\sqrt{\frac{\frac{1}{3}\,q^2 + 4\,s}{3}} \right)} \right\},$$

$$x_{,,,} = \frac{1}{2}\left\{ \sqrt{\left(-\frac{2\,q}{3} \mp 2\sqrt{\frac{\frac{1}{3}\,q^2 + 4\,s}{3}} \right)} \right\},$$

$$x_{,,,,} = \frac{1}{2}\left\{ 2\sqrt{\left(-\frac{2\,q}{3} \pm \sqrt{\frac{\frac{1}{3}\,q^2 + 4\,s}{3}} \right)} \right.$$

$$\left. -\sqrt{\left(-\frac{2\,q}{3} \mp 2\sqrt{\frac{\frac{1}{3}\,q^2 + 4\,s}{3}} \right)} \right\}.$$

PROPOSITION VI.

(87.) *When* $r = \pm\sqrt{\left(\dfrac{-2\,q^3 + 72\,q\,s \mp \frac{1}{2}\,(2\,q^2 + 24\,s)^{\frac{3}{2}}}{27} \right)}$
in the equation $x^4 + q\,x^2 + r\,x + s = 0$, *the equation will admit of a general solution.*

Demonstration.— Let the equation

$$X_{,} = x^4 + q\,x^2 \pm \sqrt{\left(\frac{-2\,q^3 + 72\,q\,s \mp \frac{1}{2}\,(2\,q^2 + 24\,s)^{\frac{3}{2}}}{27} \right)}\,x + s = 0$$

be transformed into a cubic equation by art. (77.); and let $y = \frac{y}{3}$;
then equation I., in art. (77.), becomes

$$Y_{,} = \left(y - \frac{2\,q}{3} \right)^3 + 2\,q\left(y - \frac{2\,q}{3} \right)^2 + (q^2 - 4\,s)\left(y - \frac{2\,q}{3} \right)$$

$$+ \frac{2\,q^3 - 72\,q\,s \pm \frac{1}{2}\,(2\,q^2 + 24\,s)^{\frac{3}{2}}}{27} = 0$$

Transform this into another without the second term : thus

$$Y = y^3 - (\tfrac{1}{3} q^2 + 4 s) y \pm \frac{\tfrac{1}{3} (2 q^2 + 24 s)^{\frac{3}{2}}}{27} = 0$$

Transform $Y = 0$ into the equation of differences, and we have

$$Z = z^3 - 3 (\tfrac{1}{3} q^2 + 4 s) z$$

$$\pm \sqrt{ - 4 [- (\tfrac{1}{3} q^2 + 4 s)^3] - 27 \left(\frac{\tfrac{1}{3} (2 q^2 + 24 s)^{\frac{3}{2}}}{27} \right)^2 } = 0$$

that is $z^3 - 3 (\tfrac{1}{3} q^2 + 4 s) z \pm \sqrt{ 2 (\tfrac{1}{3} q^2 + 4 s)^3 } = 0$

If the roots of $Y = 0$ are multiplied by $\sqrt{3}$, the transformed equation will become identical with $Z = 0$; therefore, by articles (54.), (57.), and (58.), the roots of $Y = 0$ will become

$$y = \pm \sqrt{ \frac{\tfrac{1}{3} q^2 + 4 s}{6 + 3 \sqrt{3}} },$$

$$y = \pm (1 + \sqrt{3}) \sqrt{ \frac{\tfrac{1}{3} q^2 + 4 s}{6 + 3 \sqrt{3}} },$$

$$y = \mp (2 + \sqrt{3}) \sqrt{ \frac{\tfrac{1}{3} q^2 + 4 s}{6 + 3 \sqrt{3}} }.$$

And the roots of equation Y, in this proposition, become (58.)

$$\left(y - \frac{2 q}{3} \right) = - \frac{2 q}{3} \pm \sqrt{ \frac{\tfrac{1}{3} q^2 + 4 s}{6 + 3 \sqrt{2}} },$$

$$\left(y - \frac{2 q}{3} \right) = - \frac{2 q}{.3} \pm (1 + \sqrt{3}) \sqrt{ \frac{\tfrac{1}{3} q^2 + 4 s}{6 + 3 \sqrt{3}} },$$

$$\left(y - \frac{2 q}{3} \right) = - \frac{2 q}{3} \mp (2 + \sqrt{3}) \sqrt{ \frac{\tfrac{1}{3} q^2 + 4 s}{6 + 3 \sqrt{3}} }.$$

And the four roots of $X_{,}$, by art. (78.), become

$$x_{,}=\frac{1}{2}\left\{-\sqrt{\left(-\frac{2q}{3}\pm\sqrt{\frac{\frac{1}{3}q^2+4s}{6+3\sqrt{3}}}\right)}-\sqrt{\left(-\frac{2q}{3}\mp(1+\sqrt{3})\sqrt{\frac{\frac{1}{3}q^2+4s}{6+3\sqrt{3}}}\right)}\right\},$$

$$x_{,,}=\frac{1}{2}\left\{-\sqrt{\left(-\frac{2q}{3}\pm\sqrt{\frac{\frac{1}{3}q^2+4s}{6+3\sqrt{3}}}\right)}+\sqrt{\left(-\frac{2q}{3}\mp(2+\sqrt{3})\sqrt{\frac{\frac{1}{3}q^2+4s}{6+3\sqrt{3}}}\right)},$$

$$x_{,,,}=\frac{1}{2}\left\{+\sqrt{\left(-\frac{2q}{3}\pm\sqrt{\frac{\frac{1}{3}q^2+4s}{6+3\sqrt{3}}}\right)}-\sqrt{\left(-\frac{2q}{3}\mp(1+\sqrt{3})\sqrt{\frac{\frac{1}{3}q^2+4s}{6+3\sqrt{3}}}\right)},$$

$$x_{,,,,}=\frac{1}{2}\left\{+\sqrt{\left(-\frac{2q}{3}\pm\sqrt{\frac{\frac{1}{3}q^2+4s}{6+3\sqrt{3}}}\right)}+\sqrt{\left(-\frac{2q}{3}\mp(2+\sqrt{3})\sqrt{\frac{\frac{1}{3}q^2+4s}{6+3\sqrt{3}}}\right).$$

COROLLARY.

(88.) If the coefficients of $Y = 0$ in this proposition, be represented by b and c; and those of $Z = 0$ be represented by $b_{,}$ and $c_{,}$, then

$$\frac{b^3}{c^2} = \frac{b_{,}^3}{c_{,}^2} = - 13 \cdot 5$$

that is $\quad b^3 \ : \ c^2 \ :: \ b_{,}^3 \ : \ c_{,}^2 \ :: \ - 13 \cdot 5 \ : \ 1$

(89.) If the value of r in proposition v. be compared with the value of r in proposition vi., a remarkable property will at once be perceived, namely, that all the terms of the two values are alike, with the exception of the coefficient of the radical quantity, $\left(\tfrac{1}{3} q^2 + 4 s\right)^{\frac{3}{2}}$ which varies from 2 to $\sqrt{2}$: thus

Prop. v. gives

$$r = \pm \sqrt{\left(- \frac{2 q^3}{27} + \frac{8 q s}{3} \mp \frac{2 \left(\tfrac{1}{3} q^2 + 4 s\right)^{\frac{3}{2}}}{3 \sqrt{3}} \right)}$$

Prop. vi. gives

$$r = \pm \sqrt{\left(- \frac{2 q^3}{3} + \frac{8 q s}{3} \mp \frac{2^{\frac{1}{2}} \cdot \left(\tfrac{1}{3} q^2 + 4 s\right)^{\frac{3}{2}}}{3 \sqrt{3}} \right)}$$

For all the values of r, as the coefficient of the radical varies between 2 and $\sqrt{2}$, $- \dfrac{b^3}{c^2} < 13 \cdot 5$ and $- \dfrac{b_{,}^3}{c_{,}^2} > 13 . 5$: when the coefficient is 2, $Y = 0$ contains equal roots, and $- \dfrac{b^3}{c^2} = 6 \cdot 75$, and $\dfrac{b_{,}^3}{c_{,}^2} = $ infinity: when the coefficient is greater than 2, the final term of $Z = 0$ becomes imaginary; therefore $Y = 0$ has two imaginary roots: when the coefficient is less than the $\sqrt{2}$, $- \dfrac{b^3}{c^2} > 13 \cdot 5$ and $- \dfrac{b_{,}^3}{c_{,}^2} < 13 \cdot 5$.

(90.) It will be seen in the next chapter, that when $-\dfrac{b^3}{c^2} <$ $13 \cdot 5$, a root of $Z = 0$ can be numerically obtained with much less labour than if a root of $Y = 0$ were sought : and when $-\dfrac{b^3}{c^2} >$ $13 \cdot 5$, a root of $Y = 0$ can be developed with much more ease, than to develope a root of the equation of differences $Z = 0$. We have already referred to this principle in our method of general solution. See articles (59.) and (60.)

CHAPTER VIII.

(91.) As all cubic equations can, in a very simple manner, be transformed into others whose second term is absent (31.); only this latter class need be considered; for the roots of which, when found, will enable us immediately to obtain the original roots of any given complete cubic equation. In numerical solution, the great problem has heretofore been to find the number and situation of the roots and to determine the first figure. Numerous theories have been invented to accomplish these objects; many of which are very laborious and complicated. But I shall present a new method which accomplishes both of these objects in the most simple manner, and, in most of cases, by mere inspection.

(92.) When an equation whose second term is absent is proposed, take the equation of differences; if the final term of this latter equation is real, all the roots of the proposed equation will be real. articles (38.) and (40.)

PROPOSITION I.

(93.) When the roots of a cubic equation

$$Y = y^3 + b\,y + c = 0$$

are real and the ratio of $-\dfrac{b^3}{c^2}$ is not less than $13 \cdot 5$, the first figure of the quotient, arising from the division of c by b will generally be the first figure of one of its roots, and can never be in error only in being, in some rare cases, the fraction of a unit too small.

Demonstration.—The most unfavourable case which can occur, is when the ratio of $-\dfrac{b^3}{c^2}$ is as small as $13 \cdot 5$, and when the first figure of the root is some high number, say 9. Let an equation of this description be selected; for example

$$Y = y^3 - 1014\,y + 8788 = 0$$

The ratio $\dfrac{(1014)^3}{(8788)^2} = 13 \cdot 5$; and the first three figures of one of its roots are $9 \cdot 51$. Without regard to the signs, divide 8788 by 1014; thus

$$1014\;)\;8788\;(\;8\,\tfrac{676}{1014}$$
$$\underline{8112}$$
$$676$$

Thus it is seen that the first quotient figure is too small by the fraction of a unit.

To determine with certainty when the quotient figure is too small, increase it by unity, and then take the square which subtract from the divisor, and see how many times this diminished divisor is contained in the dividend; thus

$8 + 1 = 9 \;\therefore\; 9^2 = 81$, and $1014 - 81 = 933 =$ diminished divisor; therefore

$$933\;)\;8788\;(\;9$$
$$8397$$

consequently 933 is the true divisor and 9 is the first figure of the root; if the dividend had not contained this diminished divisor 9 times, then it would be known that the former figure 8 was the first figure of the root. This can usually be ascertained by mere inspection.

(94.) When $-\dfrac{b^3}{c^2}$ is less than $13 \cdot 5$ and greater than $6 \cdot 75$, the roots will be real; but the roots of the equation of differences should be sought; for $-\dfrac{b_{\prime}^{\,3}}{c_{\prime}^{\,2}}$ will be greater than $13 \cdot 5$. See articles (89.) and (90.)

(95.) When $-\dfrac{b^3}{c^2} = 6 \cdot 75$ two roots of the equation will be equal.

(96.) When $-\dfrac{b^3}{c^2} < 6 \cdot 75$ the equation will have two imaginary roots.

(97.) When $-\dfrac{b^3}{c^2}$ or $-\dfrac{b_{\prime}^{3}}{c_{\prime}^{2}} = 13 \cdot 5$, then $\dfrac{z}{y} = \sqrt{8} = 1 \cdot 732 \ldots$; see art. (57.); therefore $z = 1 \cdot 732 \ldots y$; but z is the difference between two roots of Y; therefore two roots of Y cannot approximate each other, so as to have their first figures alike when both occupy the same place in the numeral scale of units, tens, &c., or tenths, hundredths, &c.; therefore, the serious difficulties connected with the old methods, when the roots approach equality, are, by this new method, entirely obviated.

Thus this new method arms us with a threefold advantage over the old:—First, in determining the character of the roots, whether real or imaginary; Second, in finding directly as a quotient figure the first figure of the root; and Third, the advantage of knowing that no other root can have this same figure, as an initial figure, when of the same denomination in the numeral scale.

(98.) The first figure of a root being found as in art. (93.), the other figures may be developed by HORNER's method, or by any other similar method. We will first give an example by HORNER's method, finding the first figure by the process given above.

EXAMPLE 1.

Required a root of the equation

$$y^3 - 21\,y + 7 = 0$$

to three places of figures.

In this $-\dfrac{b^3}{c^2} > 13 \cdot 5$, therefore a root of Y must be found, and not a root of Z.

$$
\begin{array}{lll}
 & b & c \\
1 + 0\cdot & -21\cdot00 & +7\cdot000\,(\,.3 \\
\quad\cdot3 & \cdot09 & -6\cdot273 \\
\hline
\quad\cdot3 & -20\cdot91 & \\
\quad\cdot3 & \cdot18 & \\
\hline
\quad\cdot6 & & \\
\quad\cdot3 & & \\
\hline
\quad\cdot90 & -20\cdot7300 \;^{*}+ & \cdot727000\,(\,\cdot03 \\
\quad\cdot03 & 279 \quad - & \cdot621063 \\
\hline
\quad\cdot93 & -20\cdot7021\; \dagger & \\
\quad3 & 288 & \\
\hline
\quad\cdot96 & & \\
\quad3 & & \\
\hline
\quad\cdot990 & -20\cdot673300\;^{*}+ & \cdot105937000\,(\,.005 \\
\quad5 & 4975 \quad - & \cdot103341625 \\
\hline
\quad\cdot995 & -20\cdot668325\; \dagger & \\
\end{array}
$$

The process followed here will be plain by reference to the examples. Art. (30.)

The first figure is obtained by dividing c by b, and will be of the same sign as c. The root is diminished by this figure, and the coefficients of the first transformed equation are

$$\cdot9 \qquad -20\cdot73 \qquad +\cdot727$$

one cypher is added to the first; two to the second; and three to the third; divide the third by the second, and $\cdot03$ the second figure of the root is obtained; diminish as before, and the coefficients

$$\cdot99 \qquad -20\cdot6733 \qquad +\cdot105937$$

are obtained; annex cyphers and again divide, and the third figure of the root $\cdot005$ is obtained. And in this same manner any number of figures of the root may be developed.

The coefficients marked thus * are called trial divisors; and those marked thus $\dagger$ are called the true divisors.

The work in practice is generally arranged in a more compact form ; thus

$$
\begin{array}{llll}
1 + 0 \cdot 0 & - 21 \cdot 00 & + 7 \cdot 000 & (\,.335 \\
 \cdot 3 & \cdot 09 & 6 \cdot 273 \\[2pt]
\hline
 \cdot 3 & 20 \cdot 91 & \cdot 727000 \;* \\
 \cdot 3 & \cdot 18 & \cdot 621063 \\[2pt]
\hline
 \cdot 6 & 20 \cdot 7300 \;* & \cdot 105937000 \;* \\
 \cdot 3 & 279 & \cdot 103341625 \\[2pt]
\hline
 \cdot 90 & 20 \cdot 7021 & 2595875 \\
 3 & 288 \\[2pt]
\hline
 \cdot 93 & 20 \cdot 673300 \;* \\
 3 & 4975 \\[2pt]
\hline
 \cdot 96 & 20 \cdot 668325 \\
 3 & 5000 \\[2pt]
\hline
 \cdot 990 & 20 \cdot 663325 \\
 5 \\[2pt]
\hline
 \cdot 995 \\
 5 \\[2pt]
\hline
1 \cdot 000 \\
 5 \\[2pt]
\hline
1 \cdot 005
\end{array}
$$

(99.) This is the usual method of arrangement; but the pupil will see the great disadvantage of this form, arising from the separation of the trial divisors and true divisors far from the dividend; and it is evident that the greater the number of figures developed, the greater will be this separation; to obviate this difficulty, the author proposes a new method, which is much shorter, and will retain the divisors in their appropriate place on the left of the dividend. The method proposed is to dispense entirely with the trial divisors of HORNER, and make each true divisor a trial divisor for the following figure; finding, by a very simple formula, each true divisor from the figures of the quotient.

(100.) The new method will be better understood by an illustration from the example already given. Let r_1, r_2, r_3, r_4, &c. represent the first, second, third, &c. figures of the root.

12

The last example, $y^3 - 21\,y + 7 = 0$, being given to find, by our new method, the root y' to eighteen places of figures.

$$3\,y' = \quad | \cdot 99\,\overset{1}{|}53\,\overset{|\,1}{6}5$$
$$y' = \quad \cdot 33\,51\,25\,60\,37\,37\,88\,64\,26$$

$b = -21\cdot$			$\cdot 09$
$_1$ $\quad\cdot 09$			$_1\;\cdot\; \cdot 0\ 09$
$d_1 = \;\;20\cdot 91$	$c = \;7\cdot$		$_2\; \cdot\; \cdot 27\ 25$
$_2\quad\quad \cdot 2079$	$6\cdot 273$		$_3\; \cdot\; \cdot 495\ 01$
$d_2 = \;\;20\cdot 7021$	$c_2 = \cdot 727$		$_4\; \cdot\; \cdot 1005\ 04$
$_3\quad\quad 33775$	$\cdot 621063$		$_5\; \cdot\; \cdot 20106\ 25$
$d_3 = \;\;20\cdot 668325$	$c_3 = \cdot 105937$		$_6\; \cdot\; \cdot 502680\ |36$
$_4\quad\quad 510051$	$\cdot 103341625$		$_7\; \cdot\; \cdot 603225|0$
$d_4 = \;\;20\cdot 66322449$	$c_4 = 2595375$		$_{8\,\&\,9}\; \cdot\; 301$
$_5\quad\quad 1206264$	2066322449		$_{10}\; \cdot\; \cdot\ |7:$
$d_5 = \;\;20\cdot 6631038636$	$c_5 = 529052551$		
$_6\quad\quad 25133625$	413262077272		
$d_6 = \;\;20\cdot 663078729975$	$c_6 = 115790473728$		
$_7\quad\quad 5630075\,36$	103315393649875		
d_7	$c_7 = 12475080078125$		
$_{9\,\&\,0}\quad 3099900\,0$	123397843859940		
$\quad\quad 60624$	$c_8 = \;77236218185$		
d_8	61989217481		
$_{10}\quad\quad 24937$	$c_9 = \;15247000704$		
$\quad\quad 37$	14464150743		
$20\cdot 6|6|3|0|7|2|4|9|0$	782849961		
	619892175		
	162957786		
	144641507		
	18316279		
	16530458		
	1785821		
	1653046		
	132775		
	123978		
	8797		
	8265		
	532		
	413		
	119		
	124		

RULES.

(101.)—1. Write c at a convenient distance on the right of and two lines below b. When $-\dfrac{b^3}{c^2} > 13 \cdot 5$, $\dfrac{c}{-b}$ will give the first figure of the root, art. (98.), which write in the quotient on the right of and about three horizontal lines above c.

2. Write the square of the root figure underneath the quotient, according to its position in the numeral scale ; that is, if a whole number, place units under units, tens under tens, &c. ; if a decimal, place tenths under tenths, hundredths under hundredths, &c. This square r_1^2 will be the correction to be subtracted from b, leaving a remainder equal to the first true divisor d_1.

Let c_2, c_3, c_4, &c. represent the successive dividends ; and let the operations be performed without any reference to the signs.

3. The second figure of the root will be obtained by $\dfrac{c_2}{d_1}$, which place in the quotient ; multiply r_1 by 3, and write the product over r_1, and represent the same by 3 y' ; underneath r_1^2, and two figures to the right, put r_2^2 ; underneath r_2^2, and one figure to the left, write the product of r_2 into 3 y' : the sum of the two lower lines $+ r_1^2$ (in the third line above) repeated, will be the second correction, marked $_2$, which, when subtracted from d_1 will give the second true divisor d_2.

4. $\dfrac{c_3}{d_2}$ will give the third figure of the root or r_3 ; place this in the quotient ; multiply r_2 by 3, and place the product over r_2 in the line 3 y' : if this product is greater than 10 or 20, the 1's or 2's must be carried one figure to the left and placed over the same ; that is, be placed above the line 3 y' ; underneath r_2^2, and two figures to the right, place r_3^2 ; under r_3^2, and one figure to the left, write the product of r_3 into 3 y' : the sum of the two lower lines $+$ double r_2^2 in the third line will be the third correction, which subtracted from d_2 will give d_3.

5. $\dfrac{c_4}{d_3}$ will furnish r_4 ; the fourth correction will be found as above, &c., &c.

6. When the number of corrections has reached about one-third of the required number of figures in the root, the method may be greatly abbreviated, as follows :—

For the first abbreviated correction cut off the two right hand figures of the sum as found above ; subtract this abbreviated correction from the preceding divisor, cutting off one right hand figure from the remainder, and the same will become the abbreviated divisor.

For the second abbreviation, find another root figure as above directed ; and multiply this root figure into the line 3 y' , omitting the two right hand figures of 3 y' ; place the product underneath the lower line, and one figure to the left of the right hand figure of that line ; the sum of these two lower lines will be the second abbreviated correction, which subtract from the preceding divisor, cutting off one figure from the remainder, which will become the second abbreviated divisor.

For the third, fourth, &c., abbreviations, proceed in the same manner, cutting off in each step two figures in the line 3 y' , until the corrections become nothing, after which the balance of the root figures is obtained by contracted division.

These rules will be far better understood by reference to the example, which we will now proceed more fully to explain.

$$
\begin{array}{rl}
 & 0 \cdot \\
 & .3 \\
\hline
 & \cdot 09 \\
 & \cdot 0 \\
\text{1st correction} = & \cdot 09 \\[4pt]
 & \cdot 9 \\
 & \cdot 3\,3 \\
\hline
 & \cdot 09 \\
 & \quad 09 \\
 & \quad 27 \\
\text{2nd correction} = & \cdot 2079 \\[4pt]
 & \cdot 99 \\
 & \cdot 3\,3\,5 \\
\hline
 & \quad 09 \\
 & \quad 27\ \ 25 \\
 & \quad 495 \\
\text{3rd correction} = & \cdot 033775
\end{array}
$$

$$\cdot 9\overset{1}{9}5$$
$$\cdot 3351$$

$$
\begin{array}{r}
25 \\
495\ 01 \\
1005 \\
\hline
\end{array}
$$

4th correction $=$　$\cdot 00510051$

$$\cdot 9\overset{1}{9}53$$
$$\cdot 33512$$

$$
\begin{array}{r}
01 \\
1005\ 04 \\
20106 \\
\hline
\end{array}
$$

5th correction $=$　$\overline{1206264}$

$$\cdot 9\overset{1}{9}536$$
$$\cdot 335125$$

$$
\begin{array}{r}
04 \\
20106\ 25 \\
502680 \\
\hline
\end{array}
$$

6th correction $=$　$\cdot 000025133625$

$$\cdot 9\overset{1}{9}53\overset{1}{6}5$$
$$\cdot 3351256$$

$$
\begin{array}{r}
25 \\
502680\ 36 \\
6032250 \\
\hline
\end{array}
$$

7th correction $=$　$\cdot 000005630075|36$

$$\cdot 9\overset{1}{9}|53|\overset{1}{6}5$$
$$\cdot 335125603$$

$$
\begin{array}{r}
603225 \\
301 \\
\hline
\end{array}
$$

8th correction $=$　$\cdot 0000006062|4$

$$|\cdot 9\overset{1}{9}|53|\overset{1}{6}5$$
$$\cdot 3351256037$$

$$
\begin{array}{r}
301 \\
7 \\
\hline
\end{array}
$$

9th correction $=$　$\cdot 000000003|7$

By inspection of the group of figures in the example, underneath the quotient, it will be seen that each of the three lines whose sum gives the respective corrections in the foregoing explanations, is embraced in that small group; and that the figures in each exist in their proper position in regard to the numeral scale.

Each correction is obtained by simply squaring the last found root figure, and placing the same on the right of the last line; and then forming a new line by the product of this root figure into 3 y': thus each correction is found by a momentary process, or by simply furnishing an additional line to the group. Thus this new method saves an immense amount of labour, and introduces a simplicity almost equivalent to that of extracting the square root.

(102.) In the last example, although eighteen figures of one of the roots are developed, yet only ten or eleven figures of the divisor are rendered permanent; in finding the other roots of the equation, this permanent divisor can be used to great advantage; if, therefore, we can by some short process obtain seventeen or eighteen permanent figures, it will be desirable. We shall show how this, by the aid of the eighteen figures of the root, may be done.

Let r_m be the mth figure of the root; let d be the $\frac{m}{2}$th or $\frac{m+1}{2}$th divisor, which contains m corrected figures; let p be the permanent divisor sought; then we shall have,

when m is even,

$$\text{I.} \ldots d - \left\{ 6\left(r_1 + r_2 + \cdots + r_{\frac{m}{2}}\right)\left(r_{\frac{m}{2}+1} + r_{\frac{m}{2}+2} + \cdots + r_{m-1}\right) \right.$$
$$\left. + 3\left(r_1 + r_2 + \cdots + r_{\frac{m}{2}}\right) r_{\frac{m}{2}} \right\} = p$$

when m is odd,

$$\text{II.} \ldots d - \left\{ 6\left(r_1 + r_2 + \cdots + r_{\frac{m+1}{2}}\right)\left(r_{\frac{m+3}{2}} + r_{\frac{m+5}{2}} + \cdots + r_{m-1}\right) \right.$$
$$\left. + 3\left(r_1 + r_2 + \cdots + r_{\frac{m+1}{2}}\right) r_{\frac{m+1}{2}} \right\} = p$$

By these formulas the permanent divisor p can be depended upon to $m-1$ or m figures.

For instance, in the last example, the first divisor which contains m figures is the 7th; from which, when the necessary two cyphers are added, subtract the 8th and 9th correction, and we have the $\frac{m}{2}$th divisor.

	20 · 6630730998996400
8th and 9th correction	− 6062418504
$d = \frac{m}{2}$th divisor	= 20 · 6630724936577896
2nd term of formula I. =	− 44998382
$p = $ Permanent divisor =	20 · 6630724891579514

On account of the importance of this permanent divisor in obtaining the remaining roots, it would be a saving of labour to carry out, in the first column, all the figures of the corrections until the $\frac{m}{2}$th or $\frac{m+1}{2}$th divisor is reached, abbreviating in the division by cutting off the requisite number of figures, the same as if these figures were not retained.

If the student will have the patience to develope these eighteen figures of the root by HORNER's method, he will be better qualified to judge concerning the great amount of labour saved by this new method, besides the advantage of far greater simplicity of arrangement, by constantly retaining the divisors and corrections in the same horizontal columns with the dividends.

EXAMPLE 2.

(103.) Required the three roots of the equation

$$Y = y^3 \overset{b}{-} 7\, y \overset{c}{+} 7 = 0$$

to fourteen places of figures.

$-\dfrac{b^3}{c^2} = 7$; this ratio being less than $13 \cdot 5$, the equation of differences must be found ; thus

$$Z = z^3 - 3 \times 7\, z + \sqrt{-4\,(-7)^3 - 27\,(7)^2} = 0$$

that is $z^3 - 21\, z + 7 = 0$

But this equation is identical to the one in example 1 ; therefore eighteen figures of z' , which is one of the differences between two roots of Y , are already known.

Let the two remaining roots of Z be represented by z'' , $- z'''$; and let the three roots of Y be represented by y' , y'' , $- y'''$.

These five remaining roots can be found by a very simple process, explained in article (64.) ; that is, divide c by $\frac{1}{3}$th of the permanent divisor given above, and the quotient will be equal to $- y'''$; thus we shall have two roots ; the remaining four roots can be obtained directly from these : thus

$$(- 3 \cdot 0489173395223 = - y'''$$

$$9) - 20 \cdot 66307248915795 = p$$

$$- \ 2 \cdot 2|9|5|8|9|6|9|4|3|2|3|9|7|7)$$

c
7 · 00000000000000
6 · 88769082971931
——————————————
· 11230917028069
9183587772959
——————————————
2047329255110
1836717554592
——————————————
210611700518
206630724892
——————————————
8980975626
2295896943
——————————————
1685078683
1607127860
——————————————
77950823
68876908
——————————————
9073915
6887691
——————————————
2186224
2066307
——————————————
119917
114795
——————————————
5122
4592
——————————————
530
459
——————————————
71
69
——————————————
2

Therefore we have (64.)

$$\frac{c}{\frac{1}{3}p} \quad = \; - \, 3 \cdot 0489173395223 \; = \; - \, y'''$$

$$\cdot 3351256037378 \; = \; z'$$

$$\frac{y''' - z'}{2} = \quad 1 \cdot 3568958678922 \; = \; y'$$

$$y''' - y' = \quad 1 \cdot 6920214716301 \; = \; y''$$

$$y''' + y' = \quad 4 \cdot 4058132074145 \; = \; z''$$

$$- \, y''' - y'' \quad - \, 4 \cdot 7409388111524 \; = \; - \, z'''$$

(104.) This method of finding five roots of two equations after one is known is not only very simple, but far more expeditious than any other method known: to obtain the last four roots is but little more trouble than merely writing them down; and even $- \, y'''$ is obtained by contracted division, with only about one half the labour of long division. The equation $y^3 - 7\,y + 7 = 0$, of which y', y'', $- \, y'''$ are the roots, is considered one of some difficulty, being treated at some length by Lagrange; but the method which we have given has excluded all difficulties arising from the near approximation of the roots to equality.

(105.) Given $y^3 - 618246\,y + 99228483 = 0$ to find the three roots to about twelve places of figures.

The equation of differences becomes

$$z^3 - 1854738\,z + 824253709 \cdot 033335 = 0$$

<pre>
                                        1 2   · 2 1 1
                                   3 y' = 3 8 4 · 3 7 8 2
  b = -6 1 8 2 4 6 ·   y' = 1 6 8 · 1 9 6 4 2 3 4 6 3 2 8
  1   1   1 · ·   1
       6 0 8 2 4 6 ·   c = 9 9 2 2 8 4 8 3 ·   3 6
  2   4 1 6   6 0 8 2 4 6   2 · · 1 8   6 4 ·
       5 6 6 6 4 6 ·   3 8 4 0 3 8 8   3 · ·   3 8 4   · 0 1
  3   2 9 1 0 4 ·   3 3 9 9 8 7 6   4 · ·   5 0 · 4   8 1
       5 3 7 5 4 2 ·   4 4 0 5 1 2 3 · 5 · ·   4 5 · 3 8 7 |3 6
  4   4 0 1 8 · 4 1   4 3 0 0 3 3 6 · 6 · ·   3 · 0 2 7|4 2   1 6
       5 3 3 5 2 3 · 5 9   1 0 4 7 8 7 · 7 · ·   · 2 0|1 8 3 5 2
  5   9 5 · 8 1 5 1   5 3 3 5 2 · 3 5 9   8 · · ·   1|0 1
       5 3 3 4 2 7 · 7 7 4 9   5 1 4 3 4 · 6 4 1
  6   4 8 · 4 3 0 6|5 6   4 8 0 0 8 · 4 9 9 7 4 1
       5 3 3 3 7 9 · 3 4 4|2 4 4 3 4 2 6 · 1 4 1 2 5 9
  7   3 · 2 2 9|3 2 7 3 6 3 2 0 0 · 2 7 6 0 6 5
       5 3 3 3 7 6 · 1 1|4 9 1 6 6 4 2 2 5 · 8 6 5 1 9 4
  8   · 2 1|1 9   2 1 3 · 3 5 0 4 4 6
       5|3 3 3|7|5 · |9|0   1 2 · 5 1 4 7 4 8
                              1 0 · 6 6 7 5 1 8
                               1 · 8 4 7 2 3 0
                               1 · 6 0 0 1 2 8
                                 · 2 4 7 1 0 2
                                 · 2 1 3 3 5 0
                                   3 3 7 5 2
                                   3 2 0 0 2
                                     1 7 5 0
                                     1 6 0 0
                                       1 5 0
                                       1 0 7
                                        4 3
                                        4 2
                                         1
</pre>

It will be observed that, though fourteen figures of the root are obtained by the above process, yet only six figures are rendered permanent in the 8th divisor; let it be required to obtain a divisor with thirteen permanent figures; in this case $r_m = r_{13}$. By art. (102.), formula II., we have

$d = \frac{m+1}{2}$th divisor $\quad = -533376 \cdot 1149166 = $ 7th divisor, given

2nd term of form. II. $= \quad\quad +\cdot 2255143 \quad\quad$ above

$p =$ permanent divisor $= -533375 \cdot 8894023$

It will be perceived, by retaining the surplus figures, which are cut off in the above example, and not used in dividing, that the $\frac{m+1}{2}$th or $\frac{13+1}{2}$th divisor is already prepared for the correction to be applied by the second term of the formula. The retention of these surplus figures, in the first and last vertical columns, for merely the space of about one or two corrections, is but a momentary additional labour.

To obtain the remaining roots of $Y = 0$, divide c, by the permanent divisor p, the quotient will be equal to $-z'''$, which is one of the roots of the equation of differences; art. (65.)

Therefore we have

$$\frac{c_{,}}{p} \quad\quad = -1545 \cdot 352396707 = -z'''$$

$$168 \cdot 196423463 = y'$$

$$\frac{z''' - 3 v'}{2} = \quad 520 \cdot 381563159 = z'$$

$$z''' - z' \quad = \quad 1024 \cdot 970833548 = z''$$

$$y' + z' \quad = \quad 688 \cdot 577986622 = y''$$

$$-y'' - y' \quad = - \quad 856 \cdot 774410085 = -y'''$$

Thus not only the three required roots y', y'', $-y'''$ are obtained, but the three differences are also made known.

(106.) The student should be careful to remember that the symbols r_1, r_2, r_3, r_4, &c., not only represent the figures of the root, but also their denomination in the numeral scale; for instance, in this example,

$$r_1 = 100 \cdot$$
$$r_2 = 60 \cdot$$
$$r_3 = 8 .$$
$$r_4 = \cdot 1$$
$$r_5 = \cdot 09$$
$$r_6 = \cdot 006$$
$$r_7 = \cdot 0004$$
$$\&c. \qquad \&c.$$

In beginning the operation of development, the rules relating to the multiplication, division, &c. of decimals must be strictly observed; but when the work is once started, there is no further trouble, and the decimal point may be omitted; for the orderly arrangement of the system points out the exact place of all the figures, as may be seen by reference to the last two examples.

EXAMPLE 4.

(107.) Required the roots of the equation

$$Y = y^3 - \overset{b}{6 \cdot 75} \, y + \overset{c}{6 \cdot 749} = 0$$

to about six places of decimals.

Equation of differences $Z = z^3 - \overset{b_{/}}{20 \cdot 25} \, z + \overset{c_{/}}{\cdot 60371599} = 0$

$- \dfrac{b_{/}^{3}}{c_{/}^{2}}$ is over 20 000, therefore a root of Z may be found by formula
IV., art. (60.)

$$\sqrt{\dfrac{- b_{/}}{- \dfrac{b_{/}^{3}}{c_{/}^{2}} - 2}} = \cdot 02981444 = z'$$

and art. (64.)

$$\dfrac{1}{9} \left(\dfrac{c}{\dfrac{- 3 \, b_{/}}{- \dfrac{b_{/}^{3}}{c_{/}^{2}} - 2} + b'} \right) = - 2 \cdot 9959501 = - y'''$$

$$\frac{y''' - z'}{2} = \quad 1 \cdot 488068 = \quad y'$$

$$y''' - y' = \quad 1 \cdot 512882 = \quad y''$$

$$y''' + y' = \quad 4 \cdot 479018 = \quad z''$$

$$- y''' - y'' = - 4 \cdot 508832 = - z'''$$

EXAMPLE 5.

(108.) Required the roots of the equation

$$Y = y^3 - \overset{b}{y} - \overset{c}{\cdot 38490017} = 0$$

to eight places of figures.

Equation of differences $Z = z^3 - 3z - \cdot 0004434154 = 0$

The ratio of $- \dfrac{b_{,}^{3}}{c_{,}^{2}}$ exceeds $100\,000\,000$; hence we have by formula v., art. (60.)

$$- \frac{c_{,}}{b'} = - \quad \cdot 0001478051 = - z'$$

$$\frac{c}{\dfrac{3 z'^2 + b_{,}}{9}} = \quad 1 \cdot 15470053 \quad = \quad y'''$$

$$\frac{- y''' + z'}{2} = - \quad \cdot 57727686 \quad = - y'$$

$$- y''' + y' = - \quad \cdot 57742417 \quad = - y''$$

$$- y''' - y' = - 1 \cdot 73197689 \quad = - z''$$

$$y''' + y'' = \quad 1 \cdot 73212470 \quad = \quad z'''$$

(109.) The equation Y of this example is very difficult by the old method, because the two roots $- y'$, $- y''$, have their first three figures alike; but by our method, the nearer the roots approximate equality, the less the labour in obtaining them.

EXAMPLE 6.

(110.) Required the roots of the equation

$$Y = y^3 \overset{b}{-} 3\,y \overset{c}{-} 1 = 0$$

to eight places of figures.

Equation of differences　$Z = z^3 \overset{b,}{-} 9\,z \overset{c,}{-} 9 = 0$

The ratio of $-\dfrac{b^3}{c^2} = 27$; hence y' must be sought.

$$-3\,y' = -\cdot 9\overset{1}{\big|}\overset{2}{2}\,1$$
$$-\ y' = -\cdot 3\,4\,7\,2\,9\,6\,3\,5\,6$$

$$\begin{array}{l} \cdot 0\,9 \\ \cdot 0\ \ 1\,6 \\ 3\,6\ \ 4\,9 \\ 7\,1\,4\ \ \big|0\,4 \\ 2\,0\big|8\,2 \\ 9\big|4 \end{array}$$

$-3\,\cdot$
　$\cdot 0\,9$
$\overline{}$
$2\cdot 9\,1$
　$.2\,1\,7\,6$
$\overline{}$
$2\cdot 6\,9\,2\,4$
　$4\,6\,3\,8\,9$
$\overline{}$
$2\cdot 6\,4\,6\,0\,1\,1$
　$7\,4\,4\,6\big|2\,4$
$\overline{}$
$2\cdot 6\,3\,8\,5\,6\big|4\,7\,6$
　$3\,0\big|2$
$\overline{}$
$2\big|\cdot 6\big|3\big|8\big|2\big|6$

$-1\,\cdot$
　$\cdot 8\,7\,3$
$\overline{}$
　$\cdot 1\,2\,7$
　$\cdot 1\,0\,7\,6\,9\,6$
$\overline{}$
　$1\,9\,3\,0\,4$
　$1\,8\,5\,2\,2\,0\,7\,7$
$\overline{}$
　$7\,8\,1\,9\,2\,3$
　$5\,2\,7\,7\,1\,3$
$\overline{}$
　$2\,5\,4\,2\,1\,0$
　$2\,3\,7\,4\,4\,3$
$\overline{}$
　$1\,6\,7\,6\,7$
　$1\,5\,8\,2\,9$
$\overline{}$
　$9\,3\,8$
　$7\,9\,1$
$\overline{}$
　$1\,4\,7$
　$1\,3\,2$
$\overline{}$
　$1\,5$
　$1\,6$
$\overline{}$

Find a permanent divisor to eight figures; that is, $m = 8$; $d = \frac{m}{2}$-th divisor $= -2 \cdot 6385647 = $ 4th divisor.

2nd term of formula I. $= \qquad + 4090$

$p = $ permanent divisor $= -2 \cdot 6381557$

Therefore we shall have (65.)

$$- \quad \cdot 3472963 = -y'$$

$$\frac{c}{p} = \frac{c_,}{-2 \cdot 6381557} = \quad 3 \cdot 4114741 = \quad z''$$

$$\frac{-z''' + 3\,y'}{2} = -1 \cdot 1847926 = -z'$$

$$-z''' + z' \quad = -2 \cdot 2266815 = -z''$$

$$-y' \quad - z' \quad = -1 \cdot 5320889 = -y''$$

$$y' \quad + y'' \quad = \quad 1 \cdot 8793852 = \quad y'''$$

EXAMPLE 7.

(111.) $y^3 - 13 \cdot 5\,y + 13 \cdot 5 = 0$

$3\,y' = \begin{vmatrix} 2 & 2 & 2 \\ 3 & 0 & 7 & 4 & 0 & 1 \end{vmatrix}$

$y' = 1 \cdot 098076211353316$

```
                                                            1 ·
 - 13 · 5
    1 ·
 ─────────                13 · 5                      · 0 0 8 1
 1 2 · 5                  1 2 · 5                      · 2 7 0  6 4
 2 · 2 7 8 1             ─────────                     2 6 1 6  0 0 4 9
 ─────────               1 ·                            2 3 0 5 8 0  |3 6
 1 0 · 2 2 1 9            · 9 1 9 9 7 1                 ─────────────
   · 3 1 2 4 2 4         ─────────────                  1 9 7 6 5 |2 6
 ─────────────           8 0 0 2 9                       6 5 |8 8
 9 · 9 0 9 4 7 6         7 9 2 7 5 8 0 8               ───────────
   2 6 5 1 8 5 8 4 9     ─────────────                   |3 |3
 ─────────────────       7 5 3 1 9 2                      |0
 9 · 8 8 2 9 5 7 4 1 5 1  6 9 1 8 0 7 0 1 9 0 5 7
   2 5 0 3 5 5 0 |9 6    ─────────────────
 ─────────────────────   6 1 3 8 4 9 8 0 9 4 3
 9 · 8 8 2 7 0 7 0 6 0 |0 5 9 2 9 6 2 4 2 3 6 0
   2 0 4 2 4 |0          ─────────────────────
 ─────────────────       2 0 8 8 7 3 8 5 8 3
   6 8 6 6 3 |6           1 9 7 6 5 3 7 3 2 7
   6 9 |2               ─────────────────────
 ─────────────           1 1 2 2 0 1 2 5 6
   6 8 5 9 |4            9 8 8 2 6 8 5 9
       |3              ─────────────────
 ─────────────────      1 3 3 7 4 3 9 7
 9 · |8 |8 |2 |6 |8 5 |9  9 8 8 2 6 8 6
                        ─────────────
                         3 4 9 1 7 1 1
                         2 9 6 4 8 0 6
                        ─────────────
                          5 2 6 9 0 5
                          4 9 4 1 3 4
                        ─────────────
                           3 2 7 7 1
                           2 9 6 4 8
                        ─────────────
                            3 1 2 3
                            2 9 6 5
                        ─────────────
                             1 5 8
                              9 9
                        ─────────────
                              5 9
                              5 9
                        ─────────────
                               0
```

EXAMPLE 8.

Required the three roots of the equation

$$Y = y^3 - \overset{b}{900}\, y + \overset{c}{10392 \cdot 30484} = 0$$

to fourteen places of figures.

Equation of differences $Z = z^3 - \overset{b_{\prime}}{2700}\, z + \overset{c_{\prime}}{1 \cdot 74293990395} = 0$

The ratio of $-\dfrac{b_{\prime}^{\ 3}}{c_{\prime}^{\ 2}}$ is so great that nine figures of the root z' can be obtained by dividing $c_{\prime}$ by $b_{\prime}$; (see art. (60.), formula v.) therefore, by art. (64.), we have

$$\frac{c_{\prime}}{-b_{\prime}} = \quad \cdot 000645533297 = \quad z'$$

$$\frac{c}{\dfrac{3\,z'^2 + b_{\prime}}{9}} = -34 \cdot 641016149374 = -y'''\!\!'$$

$$\frac{y''' - z'}{2} = 17 \cdot 320185308038 = \quad y'$$

$$y''' - y' = 17 \cdot 320830841336 = \quad y''$$

$$y''' + y' = 51 \cdot 961201457412 = \quad z''$$

$$-y'''\!\!' - y'' = -51 \cdot 961846990710 = -z'''$$

(112.) The first five figures of y' and y'' are alike ; in consequence of which the equation by the old method is one of extreme difficulty, requiring a vast amount of labour. Mathematicians who are acquainted with STURM's analysis will immediately recognize the great superiority of this new method, in not only dispensing with all theorems for the separation of the two roots, but dispensing with HORNER's method of development, and arriving at the root by common contracted division.

(113.) There is a certain class of equations, containing three imaginary roots of the form of $a \sqrt{-1}$; they may be known by b and $b_{\prime}$, both being real and positive; while c and $c_{\prime}$ are both imaginary : when c is positive, the equation will have one positive and two negative imaginary roots ; and when c is negative, the equation will have one negative and two positive imaginary roots.

The rule for finding these roots is to change the signs of the last two terms of the equation, excluding the radical $\sqrt{-1}$, and develope the roots as if they were real ; when found, affix the radical sign $\sqrt{-1}$ to each.

EXAMPLE 9.

(114.) Required the roots of the equation

$$Y = y^3 + \overset{b}{7} y - \overset{c}{7} \sqrt{-1} = 0$$

to four places of figures.

$$Z = z^3 + \overset{b,}{21} z - \overset{c,}{7} \sqrt{-1} = 0$$

Change the signs ; thus $y^3 - 7 y + 7 = 0$
Equation of differences $z^3 - 21 z + 7 = 0$

Find the roots of these two equations precisely as in examples 1 and 2, and affix the radical sign. The roots to four figures will be

$$\cdot 335 \sqrt{-1} = z'$$
$$- 3 \cdot 048 \sqrt{-1} = - y'''$$
$$1 \cdot 356 \sqrt{-1} = y'$$
$$1 \cdot 692 \sqrt{-1} = y''$$
$$4 \cdot 405 \sqrt{-1} = z''$$
$$- 4 \cdot 740 \sqrt{-1} = - z'''$$

EXAMPLE 10.

(115.) Required the roots of the equation

$$Y = y^3 .+ \overset{b}{y} + \overset{c}{\cdot 38490017} \sqrt{-1} = 0$$

to four places of decimals.

$$Z = z^3 + 3 z + \sqrt{- 4 (1)^3 - 27 (\cdot 38490017 \sqrt{-1})^2} = 0$$

that is $\quad z^3 + 3 z + \overset{b,}{\cdot 0004434154} \sqrt{-1} = 0$
$\qquad\qquad\qquad\qquad \overset{c,}{}$

Change the signs $y^3 - y - \cdot 38490017 = 0$

$$z^3 - 3 z - \cdot 0004434154 = 0$$

14

Find these roots as given in example 5, and affix to each $\sqrt{-1}$

$$- \;\cdot 0001 \; \sqrt{-1} = -z'$$
$$1\cdot 1547 \; \sqrt{-1} = \quad y'''$$
$$-\;\cdot 5772 \; \sqrt{-1} = -y'$$
$$-\;\cdot 5774 \; \sqrt{-1} = -y''$$
$$-\,1\cdot 7319 \; \sqrt{-1} = -z''$$
$$1\cdot 7321 \; \sqrt{-1} = \quad z'''$$

(116.) If the coefficients of the proposed equation $y^3 + b\,y + c = 0$ are real, and $-\dfrac{b^3}{c^2}$ is less than $6\cdot 75$, the equation will have two imaginary roots (96.), and one real root, whose sign will be contrary to that of the final term of the equation. See art. (20.) When this term is negative, find the first figure of the real positive root, by substituting for the unknown quantity $0, \cdot 1, \cdot 2, \cdot 3$, &c.; or $\cdot 01, \cdot 02, \cdot 03$, &c.; or $\cdot 001, \cdot 002$, &c.; or $1, 2, 3$, &c. up to 10; or $10, 20, 30$, &c. up to 100; or $100, 200, 300$, &c. up to 1000; and so on, until a number is found which, when substituted, will give a positive value of the equation: the first figure of the last number substituted which gives a negative value to the equation, will be the first figure of the positive real root. But when the final term of the equation is positive, substitute for the unknown quantity $-\cdot 1, -\cdot 2, -\cdot 3$, &c.; or $-\cdot 01, -\cdot 02, -\cdot 03$, &c.; or $-\cdot 001, -\cdot 002$, &c.; or $0, -1, -2, -3$, &c. down to -10; or $-10, -20, -30$, &c. down to -100; or $-100, -200, -300$, &c. down to -1000; and so on, until a number is found which will render the value of the equation negative: the first figure of the last number which renders the value of the equation positive, will be the first figure of the negative real root.

After the first figure is thus found, proceed to develope the remainder of the root, according to HORNER's method, art. (98.), example 1, or according to the abridged method given hereafter in art. (121.)

EXAMPLE 11.

(117.) Required the positive root of the equation
$$Y = y^3 - 2y - 5 = 0$$
to sixteen places of figures.

The ratio of $-\dfrac{b^3}{c^2} = \dfrac{8}{25}$; this being less than $6 \cdot 75$, the equation must contain two imaginary roots, (96.); the final sign being negative, the real root will be positive (20.); substitute 0, 1, 2, &c. As the value of the equation is positive when 3 is substituted, and negative when 2 is substituted; therefore 2 must be the first figure of the positive root.

Having found the first figure of the root, we shall proceed to develope the root to sixteen places, by the method of HORNER, introducing in the course of the operation such abbreviations, as are the most usual in working his method.

```
1     0              -2             -5            (2 · 09455
      2              4              4
     ‾2‾            ‾2‾            -1000000
      2              8              949329
     ‾4‾            100000         -50671000
      2              5481           44517584
     600            105481         -6153416000
       9             5562          5578824625
     609            11104300       -574591375000
       9             25096         558055246375
     618            11129396       -16536128625
       9             25112
     6270           115450800
        4            314125
     6274           1115764925
        4            314150
     6278           111607907500
        4            3141775
     62820          111611049275
         5           3141800
     62825          111614191075
         5
     62830
         5
     628350
         5
     628355
         5
     628360
         5
     628365
```

The rule for abbreviating is as follows ; for each additional figure of the root, strike off one figure from the right of the middle column, and two figures from the right of the first column.

We will now apply, on the present example, this abbreviating process.

$$y''' = 2 \cdot 094551481542326$$

6 2|8 3|6 5	1 1 1 6 1 4 1 9 1 0 7|5	− 16536128625
	6 2 8 4	11161425391
	1 1 1 6 1 4 2 5 3 9 1	− 5374703234 *
	6 2 8 4	4464578676
	1 1 1 6 1 4 3 1 6 7|5 *	− 910129558 †
	2 5 1	892914976
	1 1 1 6 1 4 3 4 1 9	17214582
	2 5 1	11161438
	1 1 1 6 1 4 3 6 7|0 †	6053144
	5	5580719
	1 1 1 6 1 4 3 7 2	472425
	5	446457
	1 1|1|6|1|4|3|7|7	25968
		22323
		3645
		3348
		297
		223
		74
		67
		7

The student should make himself thoroughly acquainted with this abbreviating process. First, it will be perceived that six figures of the root are obtained without any abbreviation; secondly, three figures more are obtained by cutting off at each step one figure from the right of the middle column, and two figures from the right of the first column ; and lastly, seven figures more are furnished by common contracted division. By carefully working the example more practical information can be gained, than by reading any written rules.

(118.) The two imaginary roots can be obtained by depressing the equation to a quadratic, with the root already found.

(119.) If the trial divisors are compared with the true divisors, in the preceding example, it will be perceived that after two or three figures of the root are obtained, the last true divisor can be used for a trial divisor to find the following figure of the root; therefore, in all cases when the root y''' is sought, after two or three figures of the same are obtained, the trial divisors of HORNER may be dispensed with, and the labour be greatly shortened. Likewise the three vertical columns, used in the new method, can be reduced into two.

We will now give the last example, worked after the new method.

$$6 \cdot 0\overset{2}{7}\overset{1}{2}\overset{1}{5}\overset{1}{5}3\overset{1}{2}\overset{2}{4}$$

Root	$= 2 \cdot 094551481542326$	
	$-2 \cdot$	$-5 \cdot$
r_1^2	$= 4 \cdot$	$_1\ 4 \cdot$
	$\overline{2 \cdot}\ ^1$ } trial	$\overline{1 \cdot 000000}$
$2\,r_1^2 + r_2^2$	$= 8 \cdot 00$ } divisor.	$_{2\,\&\,3}\ \cdot 949329$
$3\,r_1\,r_2$	$= 0 \cdot 0$	$\overline{50671000}$
	$\overline{10 \cdot 00}\ ^2$	$_4\ 44517584$
$2\,r_2^2 + r_3^2$	$= \quad\quad 081$	$\overline{6153416000}$
$3\,(r_1 + r_2)\,r_3$	$= \quad\quad \cdot 540$	$_5\ 5578824625$
	$\overline{10 \cdot 5481}\ ^3$	$\overline{574591375000}$
$2\,r_3^2 + r_4^2$	$= \quad\quad 16216$	$_6\ 558055246375$
$3\,(r_1 + r_2 + r_3)\,r_4$	$= \quad\quad 2508$	$\overline{16536128625}$
	$\overline{11 \cdot 129396}\ ^4$	$_7\ 11161425391$
$2\,r_4^2 + r_5^2$	$= \quad\quad 3225$	$\overline{5374703234}$
$3\,(r_1 + r_2 + r_3 + r_4)\,r_5$	$= \quad\quad 31410$	$_8\ 4464573675$
	$\overline{11 \cdot 15764925}\ ^5$	$\overline{910129559}$
$2\,r_5^2 + r_6^2$	$= \quad\quad 5025$	$_9\ 892914976$
$3\,(r_1 + \,.\,.\, + r_5)\,r_6$	$= \quad\quad 314175$	$\overline{17214583}$
	$\overline{11 \cdot 1611049275}\ ^6$	$_{10}\ 11161438$
$2\,r_6^2$	$= \quad\quad 50$	$\overline{6053145}$
$3\,(r_1 + \,.\,.\, + r_6)\,r_7$	$= \quad\quad 628865$	$\overline{5580718}$
	$\overline{11 \cdot 161425391|1}\ ^7$	$\overline{472427}$
$3\,(r_1 + r_2 + r_3 + r_4)\,r_9$	$= \quad\quad 2513|4$	$\overline{446458}$
	$\overline{3418|8}\ ^8$	$\overline{25969}$
$3\,(r_1 + r_2)\,r_9$	$= \quad\quad 50|2$	$\overline{22323}$
	$\overline{372|0}\ ^9$	$\overline{3644}$
	$|0$	$\overline{3368}$
	$11 \cdot 1|6|1|4|3|7|7\ ^{10}$	$\overline{298}$
		$\overline{223}$
		$\overline{75}$
		$\overline{67}$
		$\overline{8}$

In art. (121.) a still greater abbreviation will be given, with rules for retaining the true divisors directly on the left of the respective dividends.

EXÁMPLE 11.

(120.) Required the negative root of the equation

$$Y = y^3 - 2y + 17438295482 = 0 \text{ to sixteen figures.}$$

It will immediately be perceived, that the first figure of the root must be of the denomination of thousands; by substituting -1000, -2000, -3000, it will be found that -3000 changes the value of the equation from positive to negative; therefore -2000 will be the first denomination of the root, or rather -2 will be the first figure: proceed in the development by HORNER's method.

$$-y''' = -2593 \cdot 192282680995$$

1	0	-2	$+17438295482$
	-2000	4000000	7999960000
	-2000	3999998	9438299482
	-2000	8000000	-7624999000
	-4000	11999998	1812300482
	-2000	3250000	-1748971820
	-6000	15249998	64321662
	-500	3500000	-60442851
	-6500	18749998	$3878811 \cdot 000$
	-500	683100	$-2017172 \cdot 291$
	-7000	19433098	$1861638 \cdot 709000$
	-500	691200	$-1815588 \cdot 017759$
	-7500	20124298	$46050 \cdot 621241$
	-90	23319	$-40347 \cdot 833374$
	-7590	20147617	$5702 \cdot 787867$
	-90	23328	$-4034 \cdot 786762$
	-7680	$20170945 \cdot 00$	$1668 \cdot 031105$
	-90	$777 \cdot 91$	$-1613 . 914880$
	-7770	$20171722 \cdot 91$	$54 \cdot 086225$
	-3	$777 \cdot 92$	$40 \cdot 347873$
	-7773	$20172500 \cdot 8300$	$13 \cdot 738352$
	-3	$700 \cdot 1451$	$12 \cdot 104362$
	-7776	$20173200 \cdot 9751$	$1 \cdot 633990$
	-3	$700 \cdot 1532$	$1 \cdot 613915$
	$-7779 \cdot 0$	$20173901 \cdot 1283$	20075
	$\cdot 1$	$15 \cdot 559$	18156
	$-7779 \cdot 1$	$20173916 \cdot 687$	1919
	$\cdot 1$	$15 \cdot 559$	1816
	$-7779 \cdot 2$	$20173932 \cdot 246$	103
	$\cdot 1$	$1 56$	101
	$-7779 \cdot 30$	$20173933 \cdot 81$	2
	9	$1 \cdot 55$	
	$-7779 \cdot 39$	$20173935 \cdot 36$	
	9	$\cdot 6$	
	$-7779 \cdot 48$	$20173936 \cdot 0$	
	9	$\cdot 6$	
	$-77|79 \cdot |57$	$2 0|1|7|3|9|3|6 \cdot 6$	

The other two roots of course are imaginary.

(121.) We will now give the preceding example, worked by our abridged method, which will show the amount of labour saved.

$$-3y''' = -\quad \overset{1\ \ 2}{|6\ 5|}\overset{\cdot}{7\ 9}\cdot\overset{2}{|3}\ 7$$
$$-y''' = -\quad 2\ 5\ 9\ 3\cdot 1\ 9\ 2\ 2\ 8\ 2\ 6\ 8\ 0\ 9\ 9\ 5$$

$$b = -2\ \cdot$$

$r_1^2 =$	$4|2\ 5 = r_2^2$	
$d_1 =$	$3\ 9\ 9\ 9\ 9\ 9\ 8\ \cdot$	$17438295482\ .\ = c$
$3\,y'''\ r_2 =$	$30\ \ |8\ \underline{1} = r_3^2$	$-\ 7999996$
$d_2 =$	$15\ 2\ 4\ 9\ 9\ 9\ 8\ \cdot$	94382994
$3\,y'''\ r^3 =$	$6\ 7\ 5\ \ |0\ 9\ \cdot\ = r_4^2$	76249990
$d_3 =$	$19\ 4\ 3\ 3\ 0\ 9\ 8\ \cdot$	181330048
$3\,y'''\ r_4 =$	$\overset{-}{\ }\ 2\ 3\ 3\ 1$	174897882
$d_4 =$	$20\ 1\ 4\ 7\ 6\ 1\ 7\ \cdot|0\ 1\ =\ r_5^2$	$64321662\ \cdot$
$3\,y'''\ r_5 =$	$7\ 7\ 7\ \cdot\ 9$	$60442851\ \cdot$
$d_5 =$	$20\ 1\ 7\ 1\ 7\ 2\ 2\ \cdot\ 9\ 1|8\ 1$	$3878811\ \cdot$
$3\,y'''\ r_6 =$	$7\ 0\ 0\ \cdot\ 1\ 3\ 7$	$2017172\ \cdot\ 291$
$d_6 =$	$20\ 1\ 7\ 3\ 2\ 0\ 0\ \cdot\ 9\ 7\ 5\ 1\ 0\ 4$	$1861638\ \cdot\ 709$
&c.	$1\ 5\ \cdot\ 5\ 5\ 9\ 1|4$	$1815588\ \cdot\ 087759$
	$9\ 1\ 6\ \cdot\ 6\ 8\ 7|4$	$46050\ \cdot\ 621241$
	$1\ \cdot\ 5\ 5\ 5|9$	$40347\ \cdot\ 833375$
	$9\ 3\ 3\ \cdot\ 8\ 0|2$	$5702\ \cdot\ 787866$
	$\cdot\ 6\ 2|2$	$4034\ \cdot\ 786760$
	$9\ 3\ 5\ \cdot\ 9|8$	$1668\ \cdot\ 001106$
	$|1$	$1613\ \cdot\ 914878$
	$2\ 0|1|7|8|9|3|6\ \cdot\ |6$	$54\ \cdot\ 086228$
		$40\ \cdot\ 347873$
		$13\ \cdot\ 738355$
		$12\ \cdot\ 104862$
		$1\ \cdot\ 633993$
		$1\ \cdot\ 613915$
		20078
		18156
		1922
		1816
		106
		101

5

In this example, let the two trial divisors be t_2 and t_3; then

$$r_1^2 = 4$$
$$d_1 = \underline{3999998}$$
$$t_2 = 2\,r_1^2 + d_1 = 11999998$$

$$r_2^2 = 25$$
$$3\,r_1\,r_2 = 30$$
$$d_2 = \underline{15249998}$$
$$t_3 = 2\,r_2^2 + 3\,r_1\,r_2 + d_2 = 18749998$$

RULES.

In the above process, the two trial divisors t_2 and t_3 need not be written down; for they are only to be used mentally, as suggestive of the second and third root figures.

1. t_2 is obtained by mentally adding $2\,r_1^2$ to the first true divisor, d_1.

2. t_3 is obtained by mentally adding the second true divisor d_2 to $3\,r_1\,r_2 + 2\,r_2^2$.

It is seldom, if ever, in the development of y''', that a third trial divisor, distinct from a true divisor, is required.

3. With t_2, as a mental divisor, find r_2, and place its square on the right of r_1^2, and let a vertical line be placed between them; underneath d_1, place $3\,y'''\,r_2$, the right hand figure of which being placed one digit to the left of the right hand figure of r_2^2; the sum of the three rows, doubling r_1^2, is equal to the second true divisor, d_2.

4. With t_3, as a mental divisor, find r_3, and place its square at the distance of three figures on the right of, and in the same line with, $3\,y'''\,r_2$, drawing a vertical line between them; underneath d_2, write $3\,y''\,r_3$, so that its right hand figure shall be one digit to the left of the right hand figure of r_3^2; the sum of these three rows of figures $+\ 2\,r_2^2$, will be equal to the third true divisor, d_3.

5. With d_3, as a trial divisor, find r_4, and place its square at the distance of three figures on the right of, and in the same line with, $3\,y'''\,r_3$, drawing a vertical line between them; underneath d_3, write $3\,y'''\,r_4$, so that its right hand figure shall be one digit to the left of the right hand figure of r_4^2; the sum of the last three horizontal rows of figures $+\ 2\,r_3^2$, will be equal to the fourth true divisor, d_4.

6. With d_4, as a trial divisor, find r_5, and proceed as in the previous steps : and so on.

7. When the square of the root figure (which has to be doubled when added to the three rows of figures) extends to the right of the coefficient b, it may be placed on the right of the preceding true divisor ; as in the example, r_5^2, r_6^2, r_7^2 are respectively placed on the right of d_4, d_5, d_6, and vertical lines drawn between them.

8. When the abbreviation commences, one figure on the right of each true divisor is cut off; and the abbreviated corrections are obtained by cutting off successively two figures in each step, on the right of the horizontal column above the root figures, or $-3\,y'''$, as clearly shown in the example; and finally, several of the last figures of the root are obtained by contracted division.

9. To avoid repeated multiplications of the root figures by 3, the student should be careful to remember the object of the horizontal row of figures which is placed above the root figures, and how they are successively originated by multiplying each root figure by 3, beginning on the left; and remember that when the product is over ten and twenty, to carry the 1's and 2's to the place of the preceding figure, and set them over the same ; thus, by multiplying each root figure into the preceding figures of the row or rows above, the several corrections are immediately obtained.   See the instructions upon this point in art. (101.)

Besides the vast amount of labour saved by this new method, the arrangement is much more simple, by constantly retaining the respective true divisors on the left of, and in the same horizontal line with, the respective dividends.

(122) When the proposed equation has two imaginary roots, the real root will always be of the sign opposite to that of the final term of the equation.   When this root is developed, it will be of advantage to use trial divisors, till about two or three figures are found, as in art. (121.)   The root developed by my method, when the equation of differences is used, is one of the two positive or of the two negative roots, which is numerically the smallest.   In obtaining this root, we have seen that the trial divisors of HORNER are not in any case required ; for the preceding true divisor is made the trial divisor for the following figure.

(123.) We have seen that when the ratio of $-\dfrac{b^3}{c^2}$ is less than

13.5, and greater than 6·75, the equation of differences furnishes the root: but when the coefficients are large, the labour of
finding the equation of differences is increased; in such a case, the
student can, if he chooses, always dispense with the equation of
differences; and proceed, as in art. (121.), to find the root whose
sign is opposite to that of the final term of the proposed equation,
(20.), the first figure of which he must find by trial: see arts. (116.)
and (117.) When he has found this root, which we have heretofore
designated by y''', he can then find z', one of the differences, by a
very simple process: for, if the general equation be

$$y^3 + b y + c = 0$$

one of the differences will be expressed by $z' = \pm \sqrt{\dfrac{3\,c}{y'''} - b}$

After having found y''' and z', the other roots become immediately known, as has been shown in several examples. Thus the
labour of obtaining the equation of differences is avoided. Also in
finding this root, there is no danger of encountering some other root
of the same sign; for only one root can be of the sign contrary to
that of the final term; therefore Sturm's theorem for the cubic
equation is entirely unnecessary.

(124.) When the equation of differences is dispensed with, the
formula

$$z' = \pm \sqrt{\dfrac{3\,c}{y'''} - b}$$

will always show whether the proposed equation has imaginary
roots; for when c is positive, y''' will be negative; and when c is
negative, y''' will be positive (20.) Also when b is negative, the
last term under the radical will be positive. Therefore if the coefficients of the proposed equation are real, it can have no imaginary
roots if z' is real; but when z' is imaginary, the proposed equation
will have two imaginary roots.

(125.) Another easy mental process will generally detect imaginary roots; that is, cube one or two of the first figures of b, allowing
in the mind cyphers for the rest of the figures, divide this cube by
the square of one or two of the first left hand figures of c, with the
requisite number of cyphers added, and notice if the quotient is less
than 6·75; if so, the equation most likely contains two imaginary
roots; but if the mental process cannot determine sufficiently exact,

then resort to the usual method of operation by figures.   See arts. (94.), (95.), and (96.)

(126.) We shall now give a few examples for finding y''', or the root whose sign is opposite to the final term of the equation; see art. (20), working the same by the new method, as in art (121.)

EXAMPLE 1.

Required the root $- y'''$, in the equation $y^3 - 7y + 7 = 0$, to sixteen figures.

Substitute successively for y, -1, -2, -3, -4; art. (116.) -4 changes the value of the equation from $+$ to $-$; therefore -3 is the first figure of the root.

$$- \ |9 \cdot \overset{1}{0}\overset{2}{|2}\overset{2}{4}|7\,3$$
$$- \ 3 \cdot 0\,4\,8\,9\,1\,7\,3\,3\,9\,5\,2\,2\,3\,0\,5 = -y'''$$

$b = -\ 7\cdot$
$\quad\quad 9\cdot$
$\quad\quad \overline{2\cdot 00|16}$
$\quad\quad\ \cdot 3\,6|0$
$\quad\quad \overline{20\cdot 3\,6\,1\,6|6\,4}$
$\quad\quad\quad 7\,2\,9|6$
$\quad\quad \overline{20\cdot 7\,9\,7\,8\,2\,4|8\,1}$
$\quad\quad\quad 8\,2\,2\,9|6$
$\quad\quad \overline{20\cdot 8\,7\,9\,1\,4\,2\,4\,1|0\,1}$
$\quad\quad\quad 9\,1\,4\,6|7$
$\quad\quad \overline{20\cdot 8\,8\,7\,4\,6\,5\,0\,9\,7\,1|4\,9}$
$\quad\quad\quad 6\,4\,0\,2\,7\,1|1$
$\quad\quad \overline{\quad 6\,2\,0\,5\,9\,1|4}$
$\quad\quad\quad 2\,7\,4\,4|0$
$\quad\quad \overline{\quad 6\,8\,7\,3\,6|2}$
$\quad\quad\quad 2\,7|4$
$\quad\quad \overline{\quad 6\,9\,0\,3|8}$
$\quad\quad\quad\quad |8$
$\quad \overline{20\cdot |8|8|7|6|9|0|7}$

$c = \quad 7\cdot$
$\quad\quad\ -\,6\cdot$
$\quad\quad \overline{1\cdot}$
$\quad\quad\ \cdot 8\,1\,4\,4\,6\,4$
$\quad\quad\ \cdot 1\,8\,5\,5\,3\,6$
$\quad\quad\ \cdot 1\,6\,6\,3\,8\,2\,5\,9\,2$
$\quad\quad \overline{1\,9\,1\,5\,3\,4\,0\,8}$
$\quad\quad 1\,8\,7\,9\,1\,2\,2\,8\,1\,6\,9$
$\quad\quad \overline{3\,6\,2\,1\,7\,9\,8\,3\,1}$
$\quad\quad 2\,0\,8\,8\,7\,4\,6\,5\,0\,9\,7\,1$
$\quad\quad \overline{1\,5\,3\,3\,0\,5\,1\,8\,0\,0\,2\,9}$
$\quad\quad 1\,4\,6\,2\,1\,3\,3\,4\,4\,1\,4\,0$
$\quad\quad \overline{7\,0\,9\,1\,8\,3\,5\,8\,8\,9}$
$\quad\quad 6\,2\,6\,6\,3\,0\,6\,2\,0\,9$
$\quad\quad \overline{\ \ 8\,2\,5\,5\,2\,9\,6\,8\,0}$
$\quad\quad 6\,2\,6\,6\,3\,0\,7\,1\,1$
$\quad\quad \overline{1\,9\,8\,8\,9\,8\,9\,6\,9}$
$\quad\quad 1\,8\,7\,9\,8\,9\,2\,1\,6$
$\quad\quad \overline{1\,0\,9\,0\,9\,7\,5\,3}$
$\quad\quad 1\,0\,4\,4\,3\,8\,4\,5$
$\quad\quad \overline{4\,6\,5\,9\,0\,8}$
$\quad\quad 4\,1\,7\,7\,5\,4$
$\quad\quad \overline{4\,8\,1\,5\,4}$
$\quad\quad 4\,1\,7\,7\,5$
$\quad\quad \overline{6\,3\,7\,9}$
$\quad\quad 6\,2\,6\,6$
$\quad\quad \overline{1\,1\,3}$
$\quad\ \cdot\ \ 1\,0\,4$
$\quad\quad \overline{9}$

The root $-y'''$ in this equation is found, art. (103.), by another process to fourteen places of figures.

EXAMPLE 2.

Find the root y''', in equation $y^3 - 6912\,y - 179712 = 0$, to nine places of figures.

$$|2\,7|\overset{2}{9}\overset{\cdot}{\cdot}7$$
$$93\cdot9415716 = y'''$$

$b = -\ 6\,9|1\,2\cdot$
$8\,1|0\,9\cdot$

$\overline{1188\cdot}$ $\qquad c = -\ 179712\cdot$
81 $\qquad\qquad\qquad 10692$

$\overline{18207\cdot|81}$ $\qquad\qquad\overline{72792.}$
$251\cdot|1$ $\qquad\qquad54621\cdot$

$\overline{19286\cdot91|16}$ $\qquad\qquad\overline{18171\cdot}$
$11\cdot26|8$ $\qquad\qquad17358\cdot219$

$\overline{19550\cdot9|0}$ $\qquad\qquad\overline{812\cdot781}$
$\cdot2|8$ $\qquad\qquad782\cdot036$

$\overline{19562|\cdot4}$ $\qquad\qquad\overline{30\cdot745}$
$\qquad\qquad\qquad\qquad\qquad19\cdot562$

$\overline{1|9|5|6|3\cdot}$ $\qquad\qquad\overline{11\cdot183}$
$\qquad\qquad\qquad\qquad\qquad9\cdot781$

$\qquad\qquad\qquad\qquad\qquad\overline{1\cdot402}$
$\qquad\qquad\qquad\qquad\qquad1\cdot369$

$\qquad\qquad\qquad\qquad\qquad\overline{33}$
$\qquad\qquad\qquad\qquad\qquad20$

$\qquad\qquad\qquad\qquad\qquad\overline{13}$
$\qquad\qquad\qquad\qquad\qquad12$

$\qquad\qquad\qquad\qquad\qquad\overline{1}$

EXAMPLE 3.

Required the root y''', in the equation $y^3 - 144\,y - 691\cdot199 = 0$, to ten places of figures.

By substituting 10, 20, for y, it is found that 20 changes the sign of the equation; therefore the first figure of the root is 1.

$$|3\overset{2}{9}\cdot|73$$

$$13\cdot91645595 = y'''$$

$$b = -\ 1|44\cdot$$
$$1|09\cdot$$
$$-\ 44\cdot \qquad\qquad c = -\ 691\cdot199$$
$$9 \qquad\qquad\qquad -\ 44$$
$$\overline{255\cdot|81} \qquad\qquad -\ \overline{1131\cdot}$$
$$85\cdot|1 \qquad\qquad\quad 765\cdot$$
$$\overline{398\cdot91|01} \qquad\quad \overline{366\cdot199}$$
$$\cdot41|7 \qquad\qquad\quad 359\cdot019$$
$$\overline{436\cdot0471|36} \qquad\quad \overline{7\cdot180}$$
$$\cdot2503|8 \qquad\qquad\quad 4\cdot360471$$
$$\overline{\cdot714|7} \qquad\qquad\quad \overline{2\cdot819529}$$
$$16|7 \qquad\qquad\qquad 2\cdot620288$$
$$\overline{\cdot98|2} \qquad\qquad\qquad \overline{\cdot199241}$$
$$|2 \qquad\qquad\qquad\quad \cdot174793$$
$$4|8|7|\cdot0|0 \qquad\qquad\quad \overline{24448}$$
$$21850$$
$$\overline{2598}$$
$$2185$$
$$\overline{413}$$
$$393$$
$$\overline{20}$$
$$22$$

EXAMPLE 4.

Find the root y''', in the equation

$$y^3 - 1675697859\, y - .26402295395502 = 0,$$

to twenty-three places of figures.

By making $y = 0$, the final term will be the value of the equation; to obtain a positive value, it will be necessary to substitute a number whose cube will consist of at least fourteen figures; therefore, the integral part of the root must have at least five figures; hence, substitute successively 10000 , 20000 , 30000 , 40000 , 50000; this latter changes the value of the equation from − to + ; therefore 4 is the first figure of the root.

$$1\,\overset{2}{\underset{2}{1}}\,\overset{1}{\underset{6}{2}}\,\overset{2}{\underset{8}{1}}\cdot\overset{\cdot}{\underset{7}{7}}\,\overset{2}{\underset{7}{2}}7$$

Root = $47267\cdot999873064229272620 = y'''$

$b = -1675697859\cdot$

    $16\,49$

   $-\,75697859\cdot$ $c = -\,26402295395502\cdot$

    $84\ \ 04$ $-\,302791436$

   $4013302141.$ $-\,29430209755$

    $282\ \ 36$ 28093114987

   $4979542141\cdot$ 13370947685

    $8496\ \ 49\cdot$ 9959084282

   $5016321741\cdot$ 34118634030

    99246 30097930446

   $5025817450\cdot81$ $40207035842\cdot$

    $127620\cdot9$ $35180722150\cdot$

   $5026987629\cdot7181$ $5026318692\cdot$

    $12762\cdot333$ $4524243866\cdot739$

   $5027078014\cdot571181$ $502069825\cdot261$

    $1276\cdot23573$ $452437021\cdot311399$

   $5027092053\cdot15611164$ $49632803\cdot949601$

    $113\cdot4431976$ $45243828\cdot478404999$

     $3442\cdot835201$ $4388975\cdot471196001$

      $9\cdot926280$ $4021674\cdot754268161$

     $3566\cdot20468$ $367300\cdot716927840$

       $\cdot42541$ $351896\cdot549634327$

     $3576\cdot5564$ $15404\cdot167293513$

        80 $15081\cdot280729669$

$5\,0\,2\,7\,0\,9\,3\,5\,7\,6\cdot9\,9$ $322\cdot886563844$

                      $301\cdot625614619$

                       $21\cdot260949225$

                       $20\cdot108374308$

                        $1\cdot152574917$

                        $1\cdot005418715$

                         $\cdot147156202$

                         $\cdot100541872$

                         46614330

                         45243842

                         1370488

                         1005419

                         365069

                         351896

                         13173

                         10054

                         3119

                         3016

                         108

                         101

The first eight figures of y'' are found without abbreviation ; the next four are found by abbreviating the corrections; and the last eleven, by contracted division.

It will be seen in art. (132.) that the root y''', of this example, is found by the method of differences, to twenty-two places of figures.

To develope the root y''' by HORNER's abbreviated method, would not only be far more intricate, but require in the operation more than double the number of figures.

EXAMPLE 5.

Given $z^3 - 2700\,z + 1\cdot74298990395 = 0$, to find $-z'''$ to seventeen places of figures.

By trial the first figure is -5 of the denomination of tens.

$$-\ 1\big|5\,3\cdot\big|7\,8\big|3\,4$$

$$-\ \ \ \ 51\cdot961846990707880 = \text{Root} = -z'''$$

$b = -2700\cdot$
$2501\cdot$
$-\ 200\cdot$ $c = \ \ \ +1\cdot74298990395$
$\ \ 15$ $+10000\cdot$
$4951\cdot81$ $+10001\cdot$
$137\cdot7$ $-\ 4951\cdot$
$5241\cdot5136$ $5050\cdot742$
$9\cdot342$ $4717\cdot359$
$5390\cdot175601$ $333\cdot383939$
$\cdot15588$ $323\cdot410536$
$5399\cdot6806816 4$ $9\cdot973403903$
$\cdot1247064$ $5\cdot399680681$
$5399\cdot9612700416$ $4\cdot573723222950$
$623541 6$ $4\cdot319969016032$
$5400\cdot09221314$ $\cdot253754206918$
93531 $\cdot216003688525$
$5400\cdot0993889$ 37750518393
1403 32400596303
$5400\cdot1004 59$ 5349922090
14 4860090413
$5\big|4\big|0\big|0\cdot\big|1\big|0{,}0\,6\,1$ 489831677
486009055
3822622
3780070
42552
37801
4751
4320
431
432

The root $- z'''$ is given in art. (111.) by the equation of differences, to fourteen places of figures.

EXAMPLE 6.

Given $y^3 - y - 1 = 0$, to find y''' to seven places of figures.

$$
\begin{array}{l}
3\cdot96\\
1\cdot324718 = y'''\\
b = -\ \overline{1\cdot}\\
+\ 1\cdot
\end{array}
$$

$$
\begin{array}{ll}
\overline{0}\cdot09 & c = -\ 1\cdot\\
\quad\cdot9 & \qquad 0\cdot\\
\overline{2\cdot99}04 & \qquad -\ \overline{1\cdot}\\
\quad 7|8 & \qquad\quad\cdot897\\
\overline{4\cdot1484}16 & \qquad\quad\overline{\cdot103}\\
\quad 15\,8|4 & \qquad\quad 82968\\
\overline{4\cdot243}1 & \qquad\quad \overline{20032}\\
\quad 2|8 & \qquad\quad 16972\\
\overline{4\cdot|2|6|2} & \qquad\quad \overline{3060}\\
& \qquad\quad 2983\\
& \qquad\quad \overline{77}\\
& \qquad\quad 43\\
& \qquad\quad \overline{34}\\
& \qquad\quad 34\\
& \qquad\quad \overline{0}
\end{array}
$$

EXAMPLE 7.

Given $y^3 + 2118246\,y + 7 = 0$, to find $-\,y'''$ to twenty-one places of decimals.

$$\frac{c}{-b} = -\cdot000003804620898611398 = -\,y'''$$

In this example the ratio of $-\dfrac{b^3}{c^2}$ is very great, consisting of about eighteen figures; hence, contracted division of the coefficients furnishes the root which is correct to the last figure. See art. (60.), formula v. This root is given among the following examples, developed by our new method to twenty-nine places of decimals.

(127.) When b is positive, $-\dfrac{b^3}{c^2}$ becomes negative; and when $-\dfrac{b^3}{c^2}$ is equal to, or numerically exceeds $-13\cdot5$, the first figure

of $\dfrac{c}{-b}$ will generally be the first figure of the root $\pm\, y'''$, being sometimes about a unit too great, which, however, is always detected,
when the square of the first figure is added to b in forming the first
true divisor; therefore, in all such cases $\dfrac{c}{-b}$ is the first trial divisor
for the first figure of the root.

EXAMPLES FOR PRACTICE.

(128.) Let y', y'', y''' represent the same roots as in former
examples.

1. $y^3 - 64\,y - 127 = 0$. $\qquad - y' = -\,2 \cdot 1368247613$

2. $y^3 - 25711\,y + 1015874 = 0$. $\quad y' = \quad 42 \cdot 4961625$

3. $y^3 - 10285324\,y - 8117639487 = 0$. $- y' = -848 \cdot 681114585$

4. $y^3 - 10285324\,y + 1 = 0$. $\dfrac{c}{-b} = y' = \cdot 000000097225911405416$

5. $y^3 - y - 2 = 0$. $\qquad\qquad y''' = 1 \cdot 521379707$

6. $y^3 - y - 721 = 0$. $\qquad\qquad y''' = 9 \cdot 00413032777$

7. $y^3 - 8910102\,y + 8019005012 = 0$. $- y''' = -3360 \cdot 9569$

8. $y^3 + 7\,y + 7 = 0$. $\qquad\qquad - y''' = -\, \cdot 896921999$

9. $y^3 + 21\,y + 7 = 0$. $\qquad\qquad - y''' = -\, \cdot 331597081$

10. $y^3 + y - 1 = 0$. $\qquad\qquad y''' = \quad \cdot 682328$

11. $y^3 + 13 \cdot 5\,y - 13 \cdot 5 = 0$. $\qquad y''' = \quad \cdot 938725228$

12. $y^3 + 135000\,y + 13500000 = 0$. $- y''' = -\,93 \cdot 8725228$

13. $y^3 + 432\,y - 1728 = 0$. $\qquad\quad y''' = 3 \cdot 86622425$

14. $y^3 + \cdot000432\,y + \cdot000001728 = 0$. $- y''' = - \cdot00386622425$

15. $y^3 + 2118246\,y + 7 = 0$.

$\quad - y''' = - \cdot 00000330462089861139828136667$

See example 7, art. (126.)

16. $y^3 - 2118246\,y - 7 = 0$, to 21 decimals. See art. (60.),

formula v. $- y' = \dfrac{c}{- b}$, $= - \cdot 00000330462089861139 8$

The same by development to twenty-nine decimals.

$\quad - y' = - \cdot 00000330462089861139831544031$

$- y'''$ in equation 15 agrees with $- y'$ in equation 16 to twenty-one decimals. It is evident that the greater the ratio of $- \dfrac{b^3}{c^3}$, the nearer will be the approximation of the two roots in the two equations.

When the ratio is diminished to about $13 \cdot 5$, or $- 13 \cdot 5$, the two roots in the two equations differ in their first figure by about a unit; that is, y' is numerically greater than y'''.

(129.) By this same process, the cube root can be developed with nearly as little labour, as is bestowed on the extraction of the square root. We will next present this very useful department of our theory under the title of

A NEW AND SIMPLE METHOD OF EXTRACTING THE
CUBE ROOT.

Divide the number into periods of three figures, beginning at the place of units. Find the greatest cube in the left hand period, and subtract the same, placing the first figure of the root, so found, over the place where the first vertical column is intended to be $\cdot t_2$, or the first trial divisor will be equal to $3r_1^2$.

For the other steps, proceed according to the rules in art. (121.) which are practically explained in several examples.

EXAMPLE 1.

Find the cube root of the number 673373097125

$$\overset{2\ 1}{2418}$$

Cube Root = 8765

$$t_2 = 3\,r_1^2 = \overline{19|49} = r_2^2$$

16|8
2092936
156|6
2286396|25
1814|0
230844225

673373097125
512
161373
146503
14870097
13718376
1151721125
1151721125

EXAMPLE 2.

Find the cube root of $y^3 - 967068262369 = 0$

$$\overset{2\ 2}{2744}$$

Cube Root = 9889

$$t_2 = 243|04 = r_2^2$$

21|6
2652464
235|2
2904784|81
2667|6
293110041

— 967068262369
729
238068
212192
25876262
23238272
2637990369
2637990369

EXAMPLE 3.

Extract the cube root of 2 to six places of decimals.

$$\overset{\cdot\ 1}{3\cdot 6\,5}$$

Cube root = 1.259921

3·04
|·6
3·64|25
·18|0
4·50258|1
337|5
4·7213
3|3
4·|7|5|8

2.
1.
1.
·728
.272
.225125
46875
42492
4383
4282
101
95
6
5
1

EXAMPLE 4.

Extract the cube root of 9 to sixteen figures.

$$\left|6 \overset{\cdot}{\cdot} \overset{2}{0}\right|4\,0\left|\overset{2}{0}\right|4$$

2 · 0 8 0 0 8 3 8 2 3 0 5 1 9 0 4 = Cube root.

1 2 · 0 0 6 4	9 ·
· 4 8 0	8 ·

1 2 · 4 8 6 4 0 0 0 0 6 4	1 ·
4 9 9 2 0 0	· 9 9 8 9 1 2

1 2 · 9 7 9 6 9 9 2 0 6 4 0 9	1 0 8 8
1 8 7 2 0 7 2	1 0 8 8 3 7 5 9 3 6 5 1 2

1 2 · 9 8 0 2 1 7 1 3 9 9	4 9 6 2 4 0 6 3 4 8 8
4 9 9 2 2	3 8 9 4 0 6 5 1 4 2 0

4 0 8 5 3	1 0 6 8 3 4 1 2 0 6 8
1 2 5	1 0 3 8 4 1 9 2 6 8 2

4 5 9 7	2 9 9 2 1 9 3 8 6
2	2 5 9 6 0 4 9 1 9

1 2 9 8 0 2 4 6 1	3 9 6 1 4 4 6 7
	3 8 9 4 0 7 3 8

6 7 3 7 2 9
6 4 9 0 1 2

2 4 7 1 7
1 2 9 8 0

1 1 7 3 7
1 1 6 8 2

5 5
5 2

3

EXAMPLE 5.

Extract the cube root of 8036066063043009001 to about nineteen places of figures.

$$|6\,0|0\,9|0\,0$$

Cube root $= 2\,0\,0\,3\,0\,0\,1 \cdot 0\,0\,0\,0\,0\,0\,8\,3\,0\,8\,3\,8$

<table>
<tr><td>1 2|0 0 0 0 0 9</td><td>. . . . . . .
8036066063043009001</td></tr>
<tr><td>1 8 0 0</td><td>8</td></tr>
<tr><td>1 2 0 1 8 0 0 9|0 0 0 0 0 1 .</td><td>36066063</td></tr>
<tr><td>6 0 0 9 0 0</td><td>36054027</td></tr>
<tr><td>1 2 0 3 6 0 3 3 0 0 9 0 0 1 ·|0 0</td><td>12036043009001</td></tr>
<tr><td></td><td>12036033009001</td></tr>
<tr><td>1 2|0 3 6 0 3 9 0 1 8 0 0 3 ·</td><td>10000000</td></tr>
<tr><td></td><td>9628831</td></tr>
<tr><td></td><td>371169</td></tr>
<tr><td></td><td>361081</td></tr>
<tr><td></td><td>10088</td></tr>
<tr><td></td><td>9629</td></tr>
<tr><td></td><td>459</td></tr>
<tr><td></td><td>361</td></tr>
<tr><td></td><td>98</td></tr>
<tr><td></td><td>96</td></tr>
<tr><td></td><td>2</td></tr>
</table>

EXAMPLE 6.

Extract the cube root of 912673001 to about twenty-nine places of figures.

```
 2|2 1|0 · 0|0 0|0 0|0 9
 9 7 0 · 0 0 0 0 0 0 3 5 4 2 7 0 7 3 3 5 6 5 3 0 0 0 4 5 9 5  = cube root.
```

```
2 4 3|4 9                              912673001 ·
  1 8|9                                729
─────────────                         ─────
2 6 2 3 9|0 0 · 0 0 0 0 0 0 0 0 0 0 0 9   183673
          8 7 3 0 0 0 0 0 0 0             183673
─────────────────────────             ─────
2 8'2'2 7,0|0 · |0'0|0 8 7 3 0 0 0 0 0 0 0 9|2 5   1 ·
              1 4 5 5 0 0 0 0 0 0 4|5     · 8 4 6 8 1 0 0 0 0 2 6 1 9 0 0 0 0 0 0 2 7
─────────────────────────             · 1 5 3 1 8 9 9 9 9 7 3 8 0 9 9 9 9 9 9 7 3
  1 8 9 1 5 0 0 0 0 0 3|1               · 1 4 1 1 3 5 0 0 0 0 9 4 5 7 5 0 0 0 0 1 6
  1 1 6 4 0 0 0 0 0|0                   ─────────────
─────────────────                      1 2 0 5 4 9 9 9 6 4 3 5 2 4 9 9 9 9 5 7
  2 0 4 8 6 4 0 0 0 0|4                 1 1 2 9 0 8 0 0 0 0 8 1 9 1 5 6 0 0 0 1
        5 8 2 0 0 0|0                   ─────────────
─────────────────                      7 6 4 1 9 9 6 3 5 3 3 0 4 3 9 9 5 6
  2 0 6 0 8 6 2 0 0|0                   5 6 4 5 4 0 0 0 0 4 1 2 1 7 2 4 0 0
        2 0 3 7 0|0                     ─────────────
─────────────────                      1 9 9 6 5 9 6 3 4 9 1 8 2 6 7 5 5 6
  2 0 6 1 6 4 7|7|0                     1 9 7 5 8 9 0 0 0 1 4 4 3 1 5 3 3 9
          2|0|                          ─────────────
─────────────────                      2 0 7 0 6 3 4 7 7 3 9 5 2 2 1 7
  2|0|6| 1|8,5|3                        1 9 7 5 8 9 0 0 0 1 4 4 3 2 9 7
                                        ─────────────
                                        9 4 7 4 4 7 7 2 5 0 8 9 2 0
                                        8 4 6 8 1 0 0 0 0 6 1 8 5 6
                                        ─────────────
                                        1 0 0 6 3 7 7 2 4 4 7 0 6 4
                                        8 4 6 8 1 0 0 0 0 6 1 8 5
                                        ─────────────
                                        1 5 9 5 6 7 2 4 4 0 8 7 9
                                        1 4 1 1 3 5 0 0 0 1 0 3 1
                                        ─────────────
                                        1 8 4 3 2 2 4 3 9 8 4 8
                                        1 6 9 3 6 2 0 0 0 1 2 4
                                        ─────────────
                                        1 4 9 6 0 4 3 9 7 2 4
                                        1 4 1 1 3 5 0 0 0 1 0
                                        ─────────────
                                        8 4 6 9 3 9 7 1 4
                                        8 4 6 8 1 0 0 0 1
                                        ─────────────
                                        1 2 9 7 1 3
                                        1 1 2 9 0 8
                                        ─────────────
                                        1 6 8 0 5
                                        1 4 1 1 3
                                        ─────────────
                                        2 6 9 2
                                        2 5 4 0
                                        ─────────────
                                        1 5 2
                                        1 4 1
                                        ─────────────
                                        1 1
```

7. Extract the cube root of − 8 , to ten places of figures.

Ans. − 1 · 442249570

8. Extract the cube root of 4 , to ten places of figures.

Ans. 1 · 587401052

9. Extract the cube root of · 0001137 , to about nine decimals.

Ans. · 048445505

10. Extract the cube root of 7 , to nine places of figures.

Ans. 1 · 91293118

11. Extract the cube root of 511 · 98 , to twelve places of figures.

Ans. 7 · 99989583198

12. Extract the cube root of 343010281732 , to seventeen places of figures.

Ans. 7000 · 0699430562381

13. Extract the cube root of 100 , to eight places of figures.

Ans. 4 · 6415888

14. Extract the cube root of · 00000000005 , to about eleven places of decimals.

Ans. · 00036840315

15. Extract the cube root of 996994017017972973 .

Ans. 998997

CHAPTER IX.

NUMERICAL SOLUTION OF BIQUADRATIC EQUATIONS.

(130.) In chapter VII. we have given a general solution of equations of the fourth degree; and also have proved that all such equations can be reduced to equations of the third degree; and have given a general cubic equation whose second term is absent, and whose coefficients are formed in terms of the coefficients of the biquadratic equation. Also a complete cubic equation is obtained from the biquadratic; and it is proved that the roots of the latter can be expressed in terms of the roots of the former. Art. (79.)

In art. (83.), these two equations, together with the equation of differences, are represented as follows:—

$$Y = y^3 - 3(q^2 + 12s)y + 72qs - 2q^3 - 27r^2 = 0$$

$$Z = z^3 - 9(q^2 + 12s)z$$

$$\pm \sqrt{-4\left(\pm 3(q^2 + 12s)\right)^3 - 27(72qs - 2q^3 - 27r^2)^2} = 0$$

$$Y_{,} = \left(\frac{y - 2q}{3}\right)^3 + 2q\left(\frac{y - 2q}{3}\right)^2 + (q^2 - 4s)\left(\frac{y - 2q}{3}\right) - r^2 = 0$$

The symbols q, r, and s are the coefficients of the general biquadratic equation.

$$X_{,} = x^4 + qx^2 + rx + s = 0$$

We shall, as heretofore, represent the roots of

$$Y \text{ by } \pm y', \pm y'', \mp y''';$$

also the roots of Z by $\pm z'$, $\pm z''$, $\mp z'''$.

It is evident that when we have found the numerical value of the roots of Y, we can substitute them in $Y_{,}$ with scarcely any labour; thus the roots of $Y_{,}$ will become known; and certain simple functions of these latter will give the four roots of the biquadratic equation. Art. (79.)

The following rules will give the whole method of operation, and will further exercise the student in the rules and formulas heretofore given.

1. Transform the proposed biquadratic equation into another without the second term. Art. (31.)

2. Substitute the numerical value of' the coefficients of the transformed biquadratic equation, in the equation Y.

3. Find by the method in art. (121.), the root $\pm\ y'''$ in equation Y. See art. (123.)

4. Find z' by formula $z' = \pm \sqrt{\dfrac{3\,c}{y'''} - b}$; arts. (123.) and (124.)

5. With y''' and z', find y' and y'', by the simple method already so often explained. Arts. (64.) and (65.); also see examples in chap. VIII.

6. Substitute y', y'', y''', in $Y_{,}$, or rather in $\dfrac{y - 2\,q}{3}$.

7. Extract the square roots of each of the three quantities $\dfrac{y' - 2\,q}{3}$, $\dfrac{y'' - 2\,q}{3}$, $\dfrac{y''' - 2\,q}{3}$, and add them together according to art. (79.), and divide by 2; the result will be the four roots of the transformed biquadratic equation, whose second term is absent.

8. Add to or subtract from each of these four roots the amount by which the roots of the original equation, by removing the second term, were diminished or increased; the result will be the required roots of the proposed complete biquadratic equation.

In case the equation of differences is used, the author's method of finding y' or z', according as the ratio of $-\dfrac{b^3}{c^2}$ or $-\dfrac{b_{_/}{}^3}{c_{_/}{}^2}$ exceeds $13 \cdot 5$, will be found much shorter and easier. It will be well for the pupil to work by both methods, and he will then be prepared to judge between the two.

We will add a few examples, leaving the plainest portions of the operation for the exercise of the student's own judgment.

EXAMPLE 1.

(131.) Required the four roots of the equation

$$x^4 - 80\,x^3 + 1998\,x^2 - 14937\,x + 5000 = 0$$

to about eleven places of figures.

Diminish the roots of the equation by 20, which is minus the fourth part of the second coefficient; this will remove the second term; and we shall have the transformed equation,

$$x^4 - 402\,x^2 + 983\,x + 25460 = 0$$

hence $q = -402,\ r = +983,\ s = +25460$

Substitute these values in Y, and we obtain

$$Y = y^3 - 1401372\,y - 633074427 = 0$$

Find by the method in art. (121.) the positive root $+ y'''$ to about eleven places of decimals; its value will be

$$+ 1365 \cdot 63115073103 = + y'''$$

Find $-z' = -\sqrt{\dfrac{3\,c}{y'''} - b} = -\quad 103 \cdot 16336781589 = -z'$

$$\frac{-y''' + z'}{2} = -\quad 631 \cdot 23389145757 = -y'$$

$$-y''' + y' = -\quad 734 \cdot 39725927346 = -y''$$

Substitute $-y'$, $-y''$, $+y'''$ in $\dfrac{y-2\,q}{3}$, and the three values will be

$$28 \cdot 20091357551$$

$$57 \cdot 58870284748$$

$$723 \cdot 21038357701$$

take the square root of each, and we shall have

$$4 \cdot 816732666$$

$$7 \cdot 588722082$$

$$26 \cdot 892571159$$

Add according to art. (79.), and divide by 2; and also add 20, according to rules 7 and 8; and the four roots of the proposed biquadratic equation will be

$$x = 34 \cdot 832280287$$

$$x = 32 \cdot 060290871$$

$$x = 12 \cdot 756441795$$

$$x = \qquad \cdot 350987047$$

Sum = second coefficient

with its sign changed $= 80 \cdot 000000000$

This result verifies the correctness of each of the roots.

EXAMPLE 2.

(132.) Required, by the author's new method, the four roots of the equation

$$X = x^4 + 312\,x^3 + 23337\,x^2 - 14874\,x + 2360 = 0 \qquad *$$

to seventeen places of decimals.

* This example is taken from a treatise, by Professor J. R. YOUNG, On the Cubic Equation. The example was prepared by Mr. Lockhart, as one of great difficulty in consequence of the near approximation of two of its roots.

Remove from $X = 0$ the second term by increasing each of the roots by the fourth part of its coefficient, or by 78; the result will be

$$X_1 = x_{\prime}^{4} - 13167\, x_{\prime}^{2} + 140970\, x_{\prime} + 32099672 = 0$$

hence $\quad q = -13167,\ r = 140970,\ s = 32099672$

Substitute these in equation $Y = 0$, art. (130.), and we have

$$Y = y^3 - \overset{b}{1675697859}\, y - \overset{c}{26402295395502} = 0$$

It will be seen by mere inspection that $-\dfrac{b^3}{c^2} < 13\cdot 5$; therefore the equation of differences should be sought, which will be as follows:

$$Z = z^3 - \overset{b_{\prime}}{5027093577}\, z - \overset{c_{\prime}}{638117\cdot 9977151561269} = 0$$

It will also be seen by mere inspection that $-\dfrac{b_{\prime}^{3}}{c_{\prime}^{2}}$ far exceeds 100000000; therefore by art. (60.), formula v., $\dfrac{c_{\prime}}{-b_{\prime}}$ will give, at least, the first twelve decimals of one of the differences; these decimals can be obtained by the expeditious process of contracted division. Let them be represented by $-a$. Also by the second term of formula i., art. (62.), nine more additional decimals will be obtained; thus

$$\frac{c_{\prime}}{-b_{\prime}} = -a = -\cdot 000126935770$$

$$-\frac{c_{\prime} + (a^2 + b_{\prime})\, a}{3\, a^2 + b_{\prime}} = -\cdot 00000000000528457805$$

$$\text{Sum} \quad = -\cdot 000126935770528457805 = -z^{\prime}$$

Thus by this simple process one of the differences is found to twenty-one places of decimals. But as we only wish to obtain the roots to seventeen places of decimals, the last four decimals of $-z^{\prime}$ may be omitted. By the aid of $-z^{\prime}$ the three roots of $Y = 0$ are quickly and very simply obtained: thus, art. (64.)

$$-\cdot 00012693577052845 = -z^{\prime}$$

$$\frac{c}{\frac{1}{3}\,(3\, z^{\prime 2} + b_{\prime})} = \quad 47267\cdot 99987306422927262 = +y^{\prime\prime\prime}$$

$$\frac{-y''' + z'}{2} = -23633 \cdot 99987306422937208 = -y'$$

$$-y''' + y' = -23633 \cdot 99999999999990054 = -y''$$

In squaring z' several of the right hand decimals may be omitted without any error: indeed, the squaring of $-a$, as above given, is abundantly sufficient to ensure exactness.

$$\text{Let } \frac{y - 2q}{3} = u, \text{ then}$$

$$u = 24533 \cdot 99995768807642421$$

$$u = 900 \cdot 00004231192354264$$

$$u = 900 \cdot 00000000000003315$$

$$\sqrt{}\, u = 156 \cdot 63332965141255792$$

$$\sqrt{}\, u = 30 \cdot 00000070519871742$$

$$\sqrt{}\, u = 30 \cdot 00000000000000055$$

Add these last three values, according to art. (79.), and we have

$$x_, = 78 \cdot 31666447310692052$$

$$x_, = 78 \cdot 31666517830563739$$

$$x_, = -48 \cdot 31666447310691997$$

$$x_, = -108 \cdot 31666517830563794$$

Subtract from each of these roots 78, and we shall obtain

$$x = \cdot 31666447310692052$$

$$x = \cdot 31666517830563739$$

$$x = -126 \cdot 31666447310691997$$

$$x = -186 \cdot 31666517830563794$$

$$\overline{\text{Sum} = -312 \cdot 00000000000000000}$$

The sum is equal, when the sign is changed, to the second coefficient of $X = 0$, verifying the correctness of each of the roots.

(133.) If the mathematician will have the patience to find the roots $-y'$, $-y''$, by STURM's analysis and HORNER's process, he will perceive the vast superiority of this new method. As these two roots do not separate until the ninth figure, the labor by the old method is uncommonly great, arising in part from the great magnitude of the coefficients.

Professor J. R. YOUNG, in his excellent work on "*The Analysis and Solution of Cubic and Biquadratic Equations*," has treated this last example according to the method of STURM and HORNER : he remarks that the example was prepared by MR. LOCKHART, and is one of considerable difficulty, being "framed expressly for the purpose of putting the modern methods and resources to a severe test."

(134.) We shall next give additional rules for the determination of the nature of the roots of a biquadratic equation, whether real or imaginary ; and also the method by which the first figure of the real roots may be found, and illustrate the same by several numerical examples.

Additional rules for the determination of the nature and situation of the roots in the biquadratic equation

$$X = x^4 + q x^2 + r x + s = 0$$

1. If q is negative and $q^2 - 4 s$ is positive (art. (77.), equation I.), and $-\dfrac{b^3}{c^2}$ is equal to or greater than $6 \cdot 75$, (arts. 94, 95, and 96,) or if b and c are each equal to nothing, * the four roots of $X = 0$ will be real: but when $-\dfrac{b^3}{c^2} < 6 \cdot 75$, two roots will be real and two imaginary. (Art. (85.), prop. IV.)

---

* If, under the circumstances stated in rule 1, $-\dfrac{b^3}{c^2} = 6 \cdot 75$, two of the four roots of $X = 0$ will be equal: and if b and c are each equal to nothing, three roots of $X = 0$ will be equal ; and their values will be expressed by $\pm \frac{1}{2} \sqrt{-\dfrac{2\,q}{3}}$

2. If $-\dfrac{b^3}{c^2} > 6 \cdot 75$, and q is positive, or if q and $q^2 - 4\,s$ are both negative, $X = 0$ will contain four imaginary roots: but if $-\dfrac{b^3}{c^2}$ is equal to † or less than $6 \cdot 75$, $X = 0$ will have two real roots, and two imaginary roots. (Art. (85.), prop. IV.)

These two rules for the determination of the nature of the roots, embrace all the cases which can possibly occur. The first figure or situation of a biquadratic root can be determined by successive substitution.

3. Substitute for the unknown quantity x in $X = 0$, according to the directions given in art. (116.), until a number is found which changes the value of the equation from positive to negative, or from negative to positive. The nature of the coefficients will, in general, indicate the denomination of the figure to be substituted, or rather, its place in the numeral scale. The number substituted, preceding the one which changes its value, will be the first root figure.

EXAMPLES.

1. Given $x^4 \overset{q}{-} 6\, x^2 \overset{r}{+} 8\, x \overset{s}{-} 3 = 0$ to determine the nature and situation of its roots.

The auxiliary cubic equation $Y = 0$, art. (130.), becomes

$$y^3 - 3 \left((-6)^2 + \overset{b}{\left(12 \times -8\right)} \right) y$$

$$+ 72 \times -6 \times -3 - \overset{c}{2} \times (-6)^3 - 27 \times 8^2 = 0$$

both b and c are equal to nothing; therefore, by rule first, the roots are all real; and by the note at the bottom of the page, three of the roots are equal; and because there are three changes of sign in the proposed equation, the three equal roots are positive; and according to the note, each is equal to $+ \frac{1}{2} \sqrt{-\dfrac{2 \times -6}{3}} = 1$, and

---

† If, under the circumstances expressed in rule 2, $-\dfrac{b^3}{c^2} = 6 \cdot 75$, the two real roots will be equal.

because the sum of the positive roots is equal to the sum of the negative roots, the remaining root must be -3.

2. Given $x^4 - \overset{q}{17} x^2 + \overset{r}{86} x - \overset{s}{20} = 0$, to find the character and situation of its roots.

The auxiliary becomes

$$y^3 - \overset{b}{147} y - \overset{c}{686} = 0$$

therefore $-\dfrac{b^3}{c^2} = -\dfrac{(-147)^3}{(-686)^2} = 6 \cdot 75$; and as q is negative, and $q^2 - 4s$ positive, the four roots, by rule first, must be real, and by the note to rule first, two of the roots must be equal; their value will be expressed by the general formula for two equal roots, art. (86.), being positive, because there are three changes of sign among the coefficients of the proposed equation; and therefore there must be three positive roots, and one negative.

The formula for the equal roots is

$$\tfrac{1}{2} \left\{ \sqrt{ \left(-\dfrac{2\,q}{3} \mp 2 \sqrt{ \dfrac{\frac{1}{3}\,q^2 + 4\,s}{3} } \right) } \right\} ; \text{ hence}$$

$$\tfrac{1}{2} \left\{ \sqrt{ \left(-\dfrac{2 \times -17}{3} + 2 \sqrt{ \dfrac{ \frac{(-17)^2}{3} + 4 \times -20 }{3} } \right) } \right\} = 2$$

The formula in art. (86.) for the two unequal roots is

$$\tfrac{1}{2} \left\{ \pm 2 \sqrt{ \left(-\dfrac{2q}{3} \pm \sqrt{ \dfrac{\frac{1}{3}q^2 + 4s}{3} } \right) } - \sqrt{ \left(-\dfrac{2\,q}{3} \mp 2\sqrt{ \dfrac{\frac{1}{3}q^2 + 4s}{3} } \right) } \right\} ;$$

hence

$$\tfrac{1}{2} \left\{ \pm 2 \sqrt{ \left(-\dfrac{2 \times -17}{3} - \sqrt{ \dfrac{ \frac{(-17)^2}{3} + 4 \times -20 }{3} } \right) } - 4 \right\} = 1, \text{ and} -5$$

3. Given $x^4 + \overset{q}{x^2} + \overset{r}{6} x + \overset{s}{4} = 0$ to find the nature and situation of the roots.

The auxiliary cubic becomes

$$y^3 \overset{b}{-} 147\,y \overset{c}{-} 686 = 0$$

This, as in the last example, gives $-\dfrac{b^3}{c^2} = 6 \cdot 75$; therefore, as q is positive, by rule second, there will be two real and two imaginary roots; and by the note to rule second, the two real roots will be equal. These may be found by the general formula, art. (86.), used in the last example, being equal to -1; the two imaginary roots may be found by the same formula, being equal to

$$1 + \sqrt{3}\,\sqrt{-1},\ 1 - \sqrt{3}\,\sqrt{-1}.$$

4. The equation $x^4 \overset{q}{-} x^2 \overset{r}{+} \cdot 6\,x \overset{s}{-} \cdot 2 = 0$ is given, to find the character of the roots, and the first figure of each real root, if such exist.

The auxiliary cubic is

$$y^3 \overset{b}{+} 4 \cdot 2\,y \overset{c}{-} 80 \cdot 8 = 0$$

As b is positive, $-\dfrac{b^3}{c^2}$ is a negative quantity, and therefore smaller than the positive quantity $6 \cdot 75$; and also q is negative and $q^2 - 4\,s$ is positive; therefore, by rule first, the proposed equation will contain two real and two imaginary roots. As the final term of the proposed equation is negative, one of the real roots must be positive, and the other negative. To find the first figure of the positive root, substitute for x, $\cdot 0$, $+ \cdot 1$, $+ \cdot 2$, &c.; the number $\cdot 8$ changes the value of the proposed equation from negative to positive; therefore, by rule third, $\cdot 7$ is the first figure of the positive root. Substitute $- 1$, $- 2$, to find the negative root; the value of the equation is changed from negative to positive by the substitution of $- 2$; therefore, by rule third, $- 1$ is the first figure of the negative root.

5. Given $x^4 - 2\,x^2 + 3\,x + 1 = 0$, to determine the nature of the roots, and find the first figure of each of the real roots, if any exist.

18

The auxiliary cubic is

$$y^3 - \overset{b}{48}\, y - \overset{c}{155} = 0$$

and $-\dfrac{b^3}{c^2} < 6 \cdot 75$; therefore, by the rules, there are two real roots, and two imaginary roots.

By substituting 0, -1, the latter changes the value from positive to negative; therefore one root must be between 0 and -1; substitute $-\cdot 1$, $-\cdot 2$, $-\cdot 3$. The sign is changed by $-\cdot 3$, therefore $-\cdot 2$ must be the first figure of one of the roots. Again substitute -1, -2, this latter restores the value of the equation from the negative to the positive; therefore -1 is the first figure of the other root.

6. $x^4 - 20\, x^2 + 60\, x - 30 = 0$. The two real roots are between the following numbers.

$(\cdot 6, \cdot 7)$; $(-5, -6)$; two roots are imaginary.

7. $x^4 - 30\, x^2 + 50\, x - 10 = 0$. $(\cdot 2, \cdot 3)$; $(1, 2)$; $(4, 5)$; $(-6, -7)$.

8. $x^4 - 1000\, x^2 + 10000\, x - 13000 = 0$. The roots are between

$(1, 2)$; $(9, 10)$; $(20, 30)$; $(-30, -40)$.

9. $x^4 - x^2 + x + 1 = 0$. One root is -1; another is between

$(-\cdot 7, -\cdot 8)$; two roots are imaginary.

10. $x^4 - 2\, x^2 - 4\, x + 8 = 0$. Four roots imaginary.

11. $x^4 + 200\, x^2 - 4000\, x + 60000 = 0$. Four roots imaginary.

12. $x^4 + 6\, x^2 - 10\, x + 8 = 0$. Four roots imaginary.

13. $x^4 - 5\, x^2 - x + 4 = 0$. The four roots are real, and are situated between the numbers.

$(\cdot 8, \cdot 9)$; $(2, 3)$; $(-1 \cdot 2, -1 \cdot 3)$; $(-1 \cdot 7, -1 \cdot 8)$.

(135.) It will be seen, by example 13, in the last art., that the two negative roots have their first figures alike; therefore two figures have to be found, in order to determine the situation of the roots: this increases the labour of substitution; and it is evident that, if the first two of the figures were alike, the difficulty would be still more increased; and if the roots should not separate, until a greater number of figures are obtained, it would be almost impossible to find by substitution the place of separation. When two roots are nega-itve and two positive, if $-\dfrac{b^3}{c^2} > 39$, the first figures of the root cannot be alike and of the same denomination; but when $-\dfrac{b^3}{c^2} < 39$, the first figures may be alike: under these circumstances, it would be advisable for the student to find the biquadratic roots by the method of the auxiliary equation of differences, according to example 2, in art. (132.)

When three roots are negative and one positive, or when three roots are positive and one negative, the first figure of the single negative or positive root can be immediately found by substitution, and the equation be depressed to a cubic, and the remaining roots be developed, according to our method for the solution of equations of the third degree.

When two roots only are real, and one is positive and one nega-tive, which will always be the case when the sign of the final term is negative, the first figure of either of the roots can be immediately obtained by substitution: but when the final term is positive, the equation must have either two positive or two negative roots, and these may approximate each other so as to have their first figures the same: in case such approximation is found to exist, as may be easily determined by a few substitutions, the student should proceed to find the roots according to the method adverted to above. Art. (132.)

(136.) Having found the first figure of a biquadratic root, accord-ing to the rules given in art. (134.), the other figures of the root may be developed to any extent required, either by the slow process of HORNER, or by a new method much more expeditious, being very similar to the one which we have devised for the development of a root in the cubic equation.

EXAMPLE.

Given $x^4 - 13167\,x^2 + 140970\,x + 32099672 = 0$, and also — 1 in the place of hundreds, as the first figure of one of its negative roots, to develope the root to about twenty places of figures. (See the roots of this equation in art. (132.))

$$4\,x = -\,40\overset{3}{2}\overset{1}{\cdot}\,\overset{\cdot}{2}\overset{2}{4}\,4$$
$$x = -\,108\cdot31666517830563794 = \text{Root.}$$

$-13167\cdot\ \ = q$	$140970\cdot = r$
$r_1^2 = 10064\cdot = r_3^2$	$h_1 r_1 = 3167$
$h_1 = -3167\cdot$	$d_1 = 457670\cdot$ } $= t_1$
$k_1 = 16833\cdot$	$k_1 r_1 = -16833$ }
$4\,x\cdot r_3 = 320 = f_3$	$h_3 r_3 = -400776\cdot$
$h_3 = 50097\cdot$	$d_2 = 1626406\cdot$ } $= t_2$
$k_3 = 53425\cdot09 = r_4^2$	$k_3 r_3 = 427400\cdot$ }
$4\,x\cdot r_4 = 129\cdot6 = f_4$	$h_4 r_4 = 17084\cdot007$
$h_4 = 56946\cdot69$	$d_3 = 2070890\cdot007$
$k_4 = 57076\cdot4701 = r_5^2$	$k_4 r_4 = 17122\cdot941$
$4\,x\cdot r_5 = 4\cdot882 = f_5$	$h_5 r_5 = 572\cdot106721$
$h_5 = 57210\cdot6721$	$d_4 = 2088585\cdot054721$
$k_5 = 57215\cdot004336 = r_6^2$	$k_5 r_5 = 572\cdot150043$
$4\,x\cdot r_6 = 2\cdot59944 = f_6$	$h_6 r_6 = 343\cdot331616456$
$h_6 = 57221\cdot986076$	$2089500\cdot5363804560$
$k_6 = 57224\cdot53558836 = r_7^2$	$k_6 r_6 = 343\cdot3472135280$
$4\,x\cdot r_7 = \cdot2599584 = f_7$	$h_7 r_7 = 34\cdot3864370570$
$h_7 = 57227\cdot395095$	$2089878\cdot2200310441$
$k_7 = 57227\cdot655054$	$k_7 r_7 = 34\cdot3365930382$
$4\,x\cdot r_8 = 25996 = f_8$	$h_8 r_8 = 3\cdot4336764661$
$h_8 = 57227\cdot94101$	$2089915\cdot99030053$
$k_8 = 57227\cdot96701$	$k_8 r_8 = 3\cdot43367802$
$4\,x\cdot r_9 = 216 = f_9$	$h_9 r_9 = \cdot28613998$
$57227\cdot9951$	$19\cdot7101185$
	$k_9 r_9 = \cdot2861400$
	$h_{10} r_{10} = 57228$
	$20\cdot0019811$
	$k_{10} r_{10} = 5723$
	$h_{11} r_{11} = 4006$
	$20\cdot01171$
	$k_{11} r_{11} = 401$
	$h_{12} r_{12} = 46$
	$20\cdot0162$
	$k_{12} r_{12} = 4$
	$h_{13} r_{13} = 0$
	$20\cdot0017$

$$
\begin{array}{l}
32099672 \cdot\ = s \\
-\ 457670 \\
\overline{-\ 18667328 \cdot} \\
\ \ 13011248 \cdot \\
\overline{-\ \ \ 656080 \cdot} \\
\ \ \ \ \ 621267 \cdot 0021 \\
\overline{\ \ \ \ \ 34812 \cdot 9979} \\
\ \ \ \ \ 20885 \cdot 85054721 \\
\overline{\ \ \ \ \ 13927 \cdot 14735279} \\
\ \ \ \ \ 12537 \cdot 0032182827 36 \\
\overline{\ \ \ \ \ 1390 \cdot 144134507264} \\
\ \ \ \ \ 1253 \cdot 926932018625 \\
\overline{\ \ \ \ \ 136 \cdot 217202488639} \\
\ \ \ \ \ 125 \cdot 394959418032 \\
\overline{\ \ \ \ \ 10 \cdot 822243070607} \\
\ \ \ \ \ 10 \cdot 449598550592 \\
\overline{\ \ \ \ \ \cdot 372644520015} \\
\ \ \ \ \ \cdot 208992000198 \\
\overline{\ \ \ \ \ \cdot 163652519817} \\
\ \ \ \ \ \cdot 146294400820 \\
\overline{\ \ \ \ \ 17358118997} \\
\ \ \ \ \ 16719360130 \\
\overline{\ \ \ \ \ 638758867} \\
\ \ \ \ \ 626976005 \\
\overline{\ \ \ \ \ 11782862} \\
\ \ \ \ \ 10449600 \\
\overline{\ \ \ \ \ 1333262} \\
\ \ \ \ \ 1253952 \\
\overline{\ \ \ \ \ 79310} \\
\ \ \ \ \ 62698 \\
\overline{\ \ \ \ \ 16612} \\
\ \ \ \ \ 14629 \\
\overline{\ \ \ \ \ 1983} \\
\ \ \ \ \ 1881 \\
\overline{\ \ \ \ \ 102} \\
\ \ \ \ \ 84 \\
\overline{\ \ \ \ \ 18}
\end{array}
$$

After the numerous examples which have been given in the chapter on the numerical solution of cubic equations, it will be unnecessary to enter into a minute explanation of the same process, applied with only a few modifications, to equations of the fourth degree, yet a few words, explanatory of the above method, may not be altogether out of place.

Let the coefficients q, r, and s be arranged in a horizontal line, at a convenient distance apart; write the first figure of the root over q in the hundreds place; take the square of $-1 = r_1$, (that is of -100) and write it under q, in the place of tens of thousands, omitting the four cyphers: multiply the algebraical sum by r_1, and place the product under the coefficient r; multiply the algebraical sum of these by r_1, and place the product under s, and take the algebraical sum.   Add h_1 to twice r_1^2 which stands above it; multiply the sum by r_1 and place the product under d_1; this product $+ d_1 = t_1 =$ the first trial divisor.

r_2 is found to be a cypher in the place of tens; for the square of r_2^2 write two cyphers on the right of r_1^2, and erect a vertical line between them; find by the trial divisor the third figure r_3 of the root; multiply the first two figures of the root by 4, and place the product directly over the root figures; place the square of r_3 (viz., 64) to the right of the two cyphers in the same line with r_1^2; multiply r_3 into the figures in the line above the root (namely 40), and place the product (namely 320) so that its right hand figure shall be one figure to the left of the right hand figure of r_3^2; then add according to the following process, or according to the following simple algebraical expressions; thus:

$$h_1 = q \; + r_1^2$$
$$k_1 = h_1 + 2\,r_1^2$$
$$h_2 = f_2 \; + r_2^2 + k_1 + 3\,r_1^2$$
$$k_2 = h_2 + f_2 + 2\,r_2^2$$
$$h_3 = f_3 \; + r_3^2 + k_2 + f_2 + 3\,r_2^2$$
$$k_3 = h_3 + f_3 + 2\,r_3^2$$
$$h_4 = f_4 \; + r_4^2 + k_3 + f_3 + 3\,r_3^2$$
$$k_4 = h_4 + f_4 + 2\,r_4^2$$

$$\cdot \qquad \cdot \qquad \cdot \qquad \cdot \qquad \cdot \qquad \cdot$$

$$h_m = f_m + r_m^2 + k_{m-1} + f_{m-1} + 3\,r_{m-1}^2$$
$$k_m = h_m + f_m + 2\,r_m^2$$

The quantities, represented by these symbols, are already arranged, according to their value in the numeral scale, and only need to be added algebraically, to obtain the required sums intended for multiplication.

While the root figures remain integral, their successive squares are retained in the line r_1^2; but when the squares of the root figures become decimals, they are attached to the right of the lines k_2, k_3, k_4, &c., as the case may be, and vertical lines drawn between them and the preceding figures. The quantities f_2, f_3, f_4, &c., are obtained by simply multiplying the last root figure, successively obtained in the line x, into the preceding figures in the line $4x$ immediately above : that is, if r_m is the last figure found, r_m multiplied into the preceding figures in the line $4x$, will give f_m: and r_m^2 should be affixed to the right of k_{m-1}

After what has been observed, art. (122.), it is scarcely necessary to repeat, that after about two trial divisors have been used mentally, they can be dispensed with, and each preceding true divisor be made the trial divisor for the following root figure.

The method of abbreviation is similar, with some modifications, to that used in the development of a root in a cubic equation : at each step one figure is cut off from the right of the middle column, and two figures cut off from the line $4x$, and from the right of the first column, for both h and k : but the method of abbreviation, as it regards the first column, will be better understood by a careful examination of the above example.   It will be perceived that seven figures of the root are obtained without any abbreviation ; six figures more by abbreviating the corrections to the true divisors in the second column ; and finally, seven figures more by contracted division.

The rule for the commencement of the abbreviation is, to begin when about one-third, or one-third plus one, of the required number of the figures of the root have been obtained.

(137.) The difference between our method and that of Horner consists, first, in the great reduction of labour ; Horner has four columns, we have but three ; the first or longest column of Horner contains about as many figures as our first two columns : and his whole process has about three times more figures than ours. Second, Horner's process, by the great inequality in the length of the four columns, separates divisors from dividends ; and parts which should be conjoined, in or near the same horizontal line, become disjointed and placed far from each other : but our method obviates this great difficulty, and introduces a greater compactness and brevity, which are considerations of no small importance in the numerical solution of equations of a higher degree than the second.

(138). In illustration of our theory, several examples will be given with the numerical operation : in some instances we shall develope the root to a great length, that the student may more fully perceive the method of abbreviation, and the facilities and expedition so happily connected with this useful process.

EXAMPLE 1.

Given $x^4 - 8 \, x^2 + 75 \, x - 10000 = 0$, and also the first figure of one of its roots, 9, to find the root to twenty-four places of figures.

$$4 \, x = 3\overset{3 \cdot 3 | 2}{6 \cdot 2 | 2 \, 4 | 0 \, 0} \qquad\qquad - 10000 \cdot$$

Root = 9 886002700947889156619312		6993·
− 3·	75·	3007·
81·	702·	2677·5616
78·	777·	329·4384
240·64	2160·	306·16628736
28·8	409·952	23·27211264
512·44	3346·952	23·261631776016
542·5264	434·016	1048·0863984
3·136	46·110592	7760877521240220352
576·3824	3827·078592	2719986462759779648
579·531236	46·362496	2716308235056603032
·23712	3·497541336	3678227703176616
582·923556	3876·938629336	3492396670229094
583·160748060004	3·498964488	185831032947522
79088·00	1166796110176·0	155217629·09895
583·3980550880	3880·438760620110176·0	30613403137627
583·3981341760	1166796268352·0	27163085216924
276808	4083787686615	3450317920703
583·398240945	3880·44033579514718·9	310435259622
583·3982·586	408378788·0	3459653·4480
	525058·5	310435259622
583·39833	744698994	35530064858
	525058	34923966707
	23336	606098151
	7452473·9	388044075
	2333	218054076
	40·8	194022037
	7452748	24032039
	4·1	23282644
	4	749395
	3880·440745279	388044
		361351

EXAMPLE 2.

The first figure of another root, in the example, art. (186.), is
− 4, in the place of tens, develope the root, by the new method,
to twenty-two figures.

$$-1\begin{vmatrix}3&1\cdot\\6&2\end{vmatrix}\cdot\begin{vmatrix}2\\2\end{vmatrix}\begin{vmatrix}2\\4\end{vmatrix}4$$

− 48·31666447310691997425 = x *

−13 1,67·	140970·	32099672·
16 64·	46268	−241460
−11567·	60365	7953672·
− 8367·	33468	7648912·
128 ·09	17784·	304760·
− 2223·	956114·	288725·8779
−815·	6520·	16034·1221
57·6	− 214·407	9621·79529279
+71.4·69	962419·593	6412·32680721
772·47 01	−231·741	5772·996980757264
1·932	− 8·322721	639·829826452736
832·2721	962179·529279	577·2963773324154864
834·204336	−8·342043	62·0334491203205136
1·15944	−·5023776456	57·7296044748890153
837·290076	962166·163459544	4·3038446454814933
838·45558836	− 5·030733528	3·8486400832859345
·11595854	− ·503838656856	·4552045621955633
839·73100476	962160·628887359144 0	·3848640068501045
839·84705388 36	−·5039082323280	703403553453693
1159598 4	− 5039847656 0	673851201171 480
839·97460093 5	96 2 1 6 0 0 74580650255	2980354 742 213
839·98620053 4	− 50399172320	28864800500081
7730 7	− 335999429 8	1028741242132
839·9985744	2082148364	962160016667
839·99934 75	−33599973 9	66581225465
7 73	− 3360000 8	57729601000
84 0·00 0 198	17125486 2	8851624465
	−336000 1	8659440150
	− 58800 0	192184315
	16730686	96216002
	−5880 0	95068313
	− 252 0	86594401
	1666937	9373912
	− 25 2	8659440
	− 8	714472
	1 6 6 6 8	673512
		40960
		38486
		2474
		1924
		550
		481
		69

* This root is given, by another process, to nineteen places of figures, in art. (132.)

EXAMPLE 3.

Given $x^4 - 402\,x^2 + 983\,x + 25460 = 0$, and also -1, in the place of tens, as the first figure of one of its roots, to develope the root to about thirteen places of figures.

$$\begin{vmatrix} 3\,2\,\cdot & 1 \end{vmatrix}$$
$$-\begin{vmatrix} 4\,6\,\cdot & 4\,6 \end{vmatrix}$$
$$-19\cdot64901295413 = \text{Root}$$

$-4{\scriptstyle|}0\,2\,\cdot = q$	$9\,8\,3 = r$	$25460 = s$
$1{\scriptstyle|}8\,1\,\cdot$	$3\,0\,2$	-40030
$-8\,0\,2\,.$	$4\,0\,0\,3$	-14570
$-1\,0\,2\,\cdot$	$1\,0\,2\,0$	$6\,5\,5\,2$
$3\,6$	$-5\,7\,5\,1$	-8018
$6\,3\,9\,\cdot$	$-\ \ 7\,2\,8$	$7\,3\,5\,7\cdot7\,8\,5\,6$
$1\,1\,6\,1\,\cdot{\scriptstyle|}3\,6$	$-1\,0\,4\,4\,9$	$6\,6\,0\cdot2\,1\,4\,4$
$4\,5\,.\,6$	$-\ \ 1\,0\,8\,5\cdot9\,7\,6$	$5\,3\,8\cdot1\,1\,9\,5\,1\,6\,1\,6$
$1\,8\,0\,9\cdot9\,6$	$1\,2\,2\,6\,2\cdot9\,7\,6$	$1\,2\,2\cdot0\,9\,4\,8\,8\,3\,8\,4$
$1\,8\,5\,6\cdot2\,8{\scriptstyle|}1\,6$	$1\,1\,1\,3\cdot7\,6\,8$	$1\,2\,1\cdot9\,1\,9\,1\,7\,6\,2\,4$
$3\cdot1\,3{\scriptstyle|}6$	$7\,6\cdot2\,4\,3\,9\,0\,4$	$\cdot1\,7\,5\,7\,0\,7\,6\,0$
$1\,9\,0\,6\cdot0\,9\,7\,6$	$1\,3\,4\,5\,2\cdot9\,8\,7\,9\,0\,4$	$\cdot1\,3\,5\,6\,3\,8\,1\,8$
$1\,9\,0\,9\cdot2\,3\,6\,6{\scriptstyle|}8\,1$	$7\,6\cdot3\,6\,9\,4\,7\,2$	$4\,0\,0\,6\,9\,4\,2$
$\cdot7\,0\,7\,0{\scriptstyle|}4$	$1\,7\cdot2\,1\,7\,7\,6\,2$	$2\,7\,1\,2\,7\,6\,8$
$1\,9\,1\,3\cdot0\,8\,4{\scriptstyle|}7$	$1\,3\,5\,4\,6\cdot5\,7\,5\,1{\scriptstyle|}3\,8$	$1\,2\,9\,4\,1\,7\,4$
$1\,9\,1\,3\cdot7\,9\,1{\scriptstyle|}9$	$1\,7\cdot2\,2\,4\,1{\scriptstyle|}2{\scriptstyle|}7$	$1\,2\,2\,0\,7\,4\,6$
${\scriptstyle|}8$	$1\,9\,1{\scriptstyle|}4{\scriptstyle|}5$	$7\,3\,4\,2\,8$
$1{\scriptstyle|}9\,1{\scriptstyle|}4\cdot5{\scriptstyle|}0\,0$	$1\,3\,5\,6\,3\cdot8\,1\,8{\scriptstyle|}4$	$6\,7\,8\,1\,9$
	$1\,9{\scriptstyle|}1$	$5\,6\,0\,9$
	$3{\scriptstyle|}8$	$5\,4\,2\,6$
	$1\,3\,5\,6\,3\cdot8\,4{\scriptstyle|}1$	$1\,8\,3$
	${\scriptstyle|}4$	$1\,3\,6$
	${\scriptstyle|}1$	$4\,7$
	$1\,3{\scriptstyle|}5\,6{\scriptstyle|}3{\scriptstyle|}\cdot8{\scriptstyle|}5$	$4\,1$
		6

EXAMPLE 4.

The first figure of one of the roots, in the equation of the last example, is − 7 in the place of units, develope the root to about eight places of figures.

```
 − 2|8 · 8
 −   7 · 2 4 3 5 5 8 2  =  Root
 − 4|0 2 ·            9 8 3             2 5 4 6 0
     |4 9 ·           2 4 7 1         − 2 4 1 7 8
 − 3 5 3 ·           3 4 5 4           1 2 8 2
 − 2 5 5 ·|0 4       1 7 8 5           1 0 5 1 · 8 9 4 4
       5 ·|6           2 0 · 4 7 2       2 3 0 · 1 0 5 6
 − 1 0 2 · 3 6       5 2 5 9 · 4 7 2     2 1 1 · 2 9 6 0
 −   9 6 · 6 8|1 6     1 9 · 3 3 6        1 9 · 8 0 9 6
       1 · 1 5|2        3 · 5 9 2         1 5 · 8 5 8 6
 −   8 9 . 8|1       5 2 8 2 · 4 0|0       2 · 9 5 1 0
 −   8 8 · 6|5          3 · 5 4|6          2 · 6 4 8 2
           9             · 2 6|2            · 3 0 7 8
 −   8 7 · 4         5 2 8 6 · 2|1          · 2 6 4 3
 −   8|7 · 3            · 2|6                4 3 5
                          |4                4 2 3
                     5|2|8|6 · |5           1 2
                                            1 1
                                           ─────
                                            1
```

(139.) 1. Required the roots of the equation

$$x^4 - x^2 + \cdot 6x - \cdot 2 = 0 \quad , \quad \text{to about seven}$$

decimals. [See example 4, art. (134.)]

Answers.
$$\begin{cases} \quad \cdot 7 4 4 9 6 3 7 \\ - \; 1 \cdot 2 6 4 6 9 8 8 \\ \text{Two roots imaginary.} \end{cases}$$

2. Required the roots of $x^4 - 2x^2 + 3x + 1 = 0$, to about eight decimals. [See art. (134.), example 5.]

Answers.
$$\begin{cases} - \quad \cdot 2 8 2 3 1 5 8 9 \\ - \; 1 \cdot 8 2 8 0 6 9 7 0 \\ \text{Two roots imaginary.} \end{cases}$$

3. Required the roots of $x^4 - 20x^2 + 60x - 80 = 0$, to about six decimals. [See art. (134.), example 6.]

$$\text{Answers.} \quad \begin{cases} \cdot 6\,2\,9\,4\,5\,4 \\ -\quad 5\cdot 6\,2\,3\,0\,8\,6 \\ \text{Two roots imaginary.} \end{cases}$$

4. Required the roots of $x^4 - 30x^3 + 50x - 10 = 0$, to about nine decimals. [See art. (134.)]

$$\text{Answers.} \quad \begin{cases} \cdot 2\,3\,2\,3\,2\,7\,3\,2\,3 \\ 1\cdot 5\,9\,1\,6\,4\,5\,5\,4\,5 \\ 4.8\,6\,7\,6\,5\,6\,0\,7\,1 \\ -\quad 6\cdot 1\,9\,1\,6\,2\,8\,9\,3\,9 \end{cases}$$

5. Required the roots of $x^4 - 1000x^2 + 10000x - 13000 = 0$, to about six decimals. [See art. (134.)]

$$\text{Answers.} \quad \begin{cases} 1\cdot 5\,8\,5\,0\,9\,7 \\ 9\cdot 4\,8\,1\,1\,3\,2 \\ 2\,4\cdot 8\,8\,1\,6\,2\,8 \\ -\,3\,5\cdot 8\,9\,7\,8\,5\,7 \end{cases}$$

(140.) The usual method of extracting the fourth root of numbers, has been, first, to extract the square root of the number, and, secondly, to extract the square root of the square root. But we propose to apply our theory to this particular class of roots, and show how they may be developed with far less labour than by the old process.

A NEW METHOD OF EXTRACTING THE FOURTH ROOT OF NUMBERS.

Divide the number into periods of four figures.

Find the greatest fourth root in the left hand period; subtract its fourth power from the same, and bring down the next period. Write the root figure, as usual, above the first vertical column; underneath this, place six times the square of the root figure: at the top of the second vertical column write four times the cube of the root figure: this will be the first trial divisor, for finding the second root figure, namely r_2; place r_2^2 on the right of $6\,r_1^2$ and draw a vertical line between them: the balance of the process is the same, as finding the roots of a biquadratic equation, so abundantly illustrated in the preceding articles.

EXAMPLE 1.

Extract the fourth root of 7971561407527201

$$
\begin{array}{ll}
& \begin{array}{l}
1\,1 \\
3\ 6\ 6\ 6
\end{array} \\
\text{Root} = \ \ 9\ 4\ 4\ 9 \\
6\,r_1^2 = 4\ 8\ 6|1\ 6 = r_1{}^? \\
\quad\quad 1\ 4|4 \\
\hline
\quad \overline{5\ 0\ 0\ 5\ 6} \\
\quad 5\ 1\ 5\ 2\ 8|1\ 6 \\
\quad\quad 1\ 5\ 0|4 \\
\hline
\quad 5\ 3\ 1\ 6\ 6\ 5\ 6 \\
\quad 5\ 3\ 3\ 1\ 7\ 2\ 8|8\ 1 \\
\quad\quad 3\ 3\ 9\ 8|4 \\
\hline
\quad 5\ 3\ 5\ 0\ 2\ 1\ 5\ 2\ 1
\end{array}
$$

$$
\begin{array}{l}
4\,r_1^3 = 2916 \\
\quad 200224 \\
\hline
\quad 8116224 \\
\quad 206112 \\
\quad 21266624 \\
\hline
\quad 3343602624 \\
\quad 21326912 \\
\quad 4815193680 \\
\hline
\quad 8369744720689
\end{array}
$$

$$
\begin{array}{l}
\quad\quad\quad\quad 7971561407527201 \\
r_1^4 = 6561 \\
\hline
\quad 14105614 \\
\quad 12464896 \\
\hline
\quad 16407180752 \\
\quad 13374410496 \\
\hline
\quad 30327702567201 \\
\quad 30327702567201 \\
\hline
\end{array}
$$

EXAMPLE 2.

Extract the fourth root of 17 to about thirteen places of figures.

$$
\begin{array}{l}
\quad 8\cdot0|3\,0 \\
\quad 2\cdot0\,3\,0\,5\,4\,3\,1\,8\,4\,8\,6\,9\ = \text{Root} \\
6\,r_1^2 = 2\,4\cdot0\,0\,0\,9 \quad 4\,r_1^3 = 3\,2\cdot \\
\quad\quad \cdot2\,4\,0 \\
\hline
\quad 2\,4\cdot2\,4\,0\,9, \\
\quad 2\,4\cdot4\,8\,2\,7|0\,0|2\,5 \\
\quad\quad 4\,0\,6\,0'0 \\
\hline
\quad 2\,4\cdot7\,2\,9\,4\,6|0 \\
\quad 2\,4\cdot7\,3\,3\,5\,2,0 \\
\quad\quad 3\,2'5 \\
\hline
\quad 2\,4\cdot7\,3\,7\,9\,1 \\
\quad 2\,4\cdot7\,3\,8|2 \\
\quad 2_{,}4\cdot7|3\,9'
\end{array}
$$

$$
\begin{array}{l}
\quad\quad\quad \cdot7\,2\,7\,2\,2\,7 \\
\quad 3\,2\cdot7\,2\,7\,2\,2\,7 \\
\quad\quad \cdot7\,3\,4\,4\,8\,1 \\
\quad 1\,2\,3\,6\,4\,7\,3\,0 \\
\hline
\quad 3\,3\cdot4\,7\,4\,0\,7\,2\,7\,3|0 \\
\quad 1\,2\,3\,6\,6\,7\,6|0 \\
\quad 9\,8\,9\,5\,1\,6 \\
\hline
\quad 3\,3\cdot4\,8\,7\,4\,2\,9\,0|1 \\
\quad 9\,8\,9\,5|3 \\
\quad 7\,4\,2|2 \\
\hline
\quad 3\,3\cdot4\,8\,8\,4\,9\,2|8 \\
\quad 7\,4|2 \\
\quad 2|5 \\
\hline
\quad 3\,3\cdot4\,8\,8\,5\,6|9 \\
\quad\quad |2 \\
\quad\quad |2 \\
\hline
\quad 3|3\cdot4|8|8|5|7
\end{array}
$$

$$
\begin{array}{l}
1.7\cdot \\
16\cdot = r_1^4 \\
\hline
1\cdot \\
\quad \cdot98181681 \\
\hline
\quad 1818819 \\
\quad 16737036365 \\
\hline
\quad 1446153635 \\
\quad 1889497160 \\
\hline
\quad 106656475 \\
\quad 100465478 \\
\hline
\quad 6190997 \\
\quad 3348857 \\
\hline
\quad 2842140 \\
\quad 2679086 \\
\hline
\quad 163054 \\
\quad 133954 \\
\hline
\quad 29100 \\
\quad 26791 \\
\hline
\quad 2309 \\
\quad 2009 \\
\hline
\quad 300 \\
\quad 301
\end{array}
$$

EXAMPLE 3.

Extract the fourth root of 90 to about thirteen places of figures.

```
                    ·  |3
            |1 2 ·|0 2                              90 ·
            3·080070288243 = Root          81· = r₁⁴
6 r₁² = 5 4·|0 0 6 4   4 r₁³ = 108·         8·
         ·|9 6 0            4·397312         8·99178496
        ─────────        ───────────          821504
        5 4·9 6 6 4       112·397312         81813502
        5 5·9 8 9 2,0 0     4·475136          336899
                 8|6            8984          233761
        ─────────          ────────          103137
        5 6·9|1 9 3        116·876432          93504
                |            898|4              9633
                              1|1               9350
                           ──────────          ─────
                           1|1|6·|8|8|0|4 3      283
                                                 234
                                                 ─────
                                                  49
                                                  46
                                                 ─────
                                                   3
                                                   3
                                                 ─────
                                                   0
```

EXAMPLE 4.

Extract the fourth root of 3 to about nine places of figures.

```
            1 ·
            4·2 4                               3·
            1·31607401 = Root.                  1· = r₁⁴
6 r₁² =     6·|0 9      4 r₁³ = 4·              2·
            1·|2               2·187            1·8561
          ─────              ────────           ·1439
            7·2 9             6·187            8889921
            8·6 7|0 1         2·601            5500079
                5|2            ·101921         5482600
          ─────────          ─────────         ───────
           10·1 9 2 1         8·889921          67479
           10·2 4 4 3|3 6      ·102443          63821
               3 1 4|4         61969            ─────
          ─────────          ─────────          3658
           10·3 2 8|1         9·054833          3347
           10·8 5 9|6         6215|7            ─────
                  |4            727              11
          ─────────          ─────────           9
           10·3 9 2          9·117|22            2̄
                                |7
                             ─────────
                             9·|1|1|8
```

EXAMPLE 5.

Extract the fourth root of 11 to about fifteen places of figures.

Ans. 1 · 82116028683787.

(141.) The great simplifications, introduced in the preceding pages, relative to the numerical solutions of equations of the Third and Fourth Degrees, will, it is hoped, render them worthy of being henceforth incorporated among the elements of Algebra, taking their place in their natural order after the quadratic, or equations of the second degree.

(142.) Equations of the higher orders, than those of the fourth degree, can be numerically solved by the aid of theorems, invented by the great mathematicians of modern times, among which may be mentioned those of BUDAN, FOURIER, and STURM. Students who are desirous of extending their enquiries beyond equations of the fourth degree, will derive great assistance from the perusal of a celebrated treatise, published by PROF. J. R. YOUNG, in 1843, entitled *"Theory and Solution of Algebraical Equations of the Higher Orders."*

D. MARPLES, PRINTER, LORD STREET, LIVERPOOL.

www.ingramcontent.com/pod-product-compliance
Lightning Source LLC
Chambersburg PA
CBHW021808190326
41518CB00007B/505